# MAKING SENSE OF JOURNALS IN THE PHYSICAL SCIENCES: FROM SPECIALTY ORIGINS TO CONTEMPORARY ASSORTMENT

Tony Stankus, MLS

## SOME ADVANCE REVIEWS

"Delves into the history of science to provide an historical perspective for a number of journals in chemistry and physics. The book provides some useful criteria for evaluating journals in these times of spiraling journal prices and decreasing materials budgets when many of us are being required to review titles for cancellation. Mr. Stankus uses a number of interesting anecdotes throughout the book which makes for very enjoyable reading."

**Gayle Baker, MS, MLS**
Reference Services Coordinator
Science & Technology
John C. Hodges Library
University of Tennessee

"Stankus is clear, practical, and unique. . . . His analyses and descriptions are people-oriented: we see how the rise of a field (and its journals) is linked to its fascinating scientists; we see that the purpose of making purchasing decisions is to help the library patron."

**Virgil Diodato, MS, PhD**
Associate Professor
School of Library and Information Science
University of Wisconsin at Milwaukee

"This book serves as a useful reminder to science librarians that scientific publications change constantly and can become more or less useful over time. Reading it could be both a reality check and self education for librarians facing major cancellations of expensive science journals."

**Sallie H. Barringer**
The Daniel E. Noble Science & Engineering Library
Arizona State University

# Making Sense of Journals in the Physical Sciences
## *From Specialty Origins to Contemporary Assortment*

# Making Sense of Journals in the Physical Sciences

## *From Specialty Origins to Contemporary Assortment*

Tony Stankus, MLS

The Haworth Press
New York • London • Norwood (Australia)

PHYSICS

*Making Sense of Journals in the Physical Sciences: From Specialty Origins to Contemporary Assortment* is monographic supplement #7 to *The Serials Librarian* (ISSN: 0897-8409). It is not supplied as part of the subscription to the journal, but is available from the publisher at an additional charge.

The Haworth Press, Inc., 10 Alice Street, Binghamton, NY 13904-1580

### Library of Congress Cataloging-in-Publication Data

Stankus, Tony.
   Making sense of journals in the physical sciences : from specialty origins to contemporary assortment / Tony Stankus.
      p.     cm. – (Monographic supplement #7 to the Serials librarian, ISSN 0897-8409)
      Includes bibliographical references and index.
      ISBN 1-56024-180-2 (alk. paper)
      1. Libraries – Special to collections – Physical sciences.   2. Physical sciences – Periodicals – Bibliography – Methodology.   3. Scientific libraries – Collection development.   4. Acquisition of scientific publications.   5. Acquisition of serial publications.   I. Title.   II. Series: Monographic supplement . . . to the Serials librarian : #7.
Z688.S3S8 1992
025.2'75002 – dc20
                                                                                    92-4092
                                                                                      CIP

# DEDICATION

To the dear friends to whom my good wife has introduced me, the greatest dowry any man could ever hope for. These are Genny Grenier, Betsy Clayborne, John and Sue Stone, and Jack and Lyn Tivnan. Their special and most necessary talent lies in keeping me from taking myself too seriously.

# CONTENTS

# ABOUT THE AUTHOR

Tony Stankus was born in Worcester, Massachusetts shortly after the arrival of his family from refugee camps in Eastern Europe. He grew up in foster homes in rural New England, graduated near the top of his high school class and won a scholarship to Holy Cross College, the Jesuit institution of Worcester. Graduating *Summa Cum Laude*, he won an assistantship to the University of Rhode Island's Graduate School of Library and Information Studies. A student paper written there developed into his first publication, followed by over 35 others in the following two decades. In 1986 and 1987, he was invited to write the annual critical overviews, "The Year's Work in Serials," for *Library Resources and Technical Services*. Haworth has published both of his prior books, *Scientific Journals: Issues in Library Selection and Management* (1987) and *Scientific Journals: Improving Library Collections Through an Analysis of Publishing Trends* (1990).

Since 1974, he has run the Science Library at Holy Cross, with the help of an assistant. All five assistants in succession have subsequently gone on to graduate training in librarianship despite having signed on with other career plans in mind initially. He has served as an editor with *Library Acquisitions: Practice and Theory, Science and Technology Libraries*, and *RQ*, where he developed a reputation for sympathetically handling nervous and uncertain first-time authors. His "Sci-Tech Collections" and "Alert Collector" columns continue to appear in each issue of the latter two journals. In 1982, the University of Rhode Island named him to their adjunct faculty as their instructor for the Special Libraries course. In 1983, Sigma Xi, the national scientific honors society, awarded him a certificate of recognition for his efforts on behalf of promoting research at Holy Cross. In 1989 he received an award for outstanding chairmanship from the working reference librarians of the Worcester Area Cooperating Libraries. In 1990, a study in *College and Research Libraries* named Tony the second most productive author nationally among 1,373 authors of articles in journals considered key to academic librarianship in the 1980s. In 1991, the ten year accreditation report of the New England Association of Secondary Schools and Colleges specifically cited his "particularly effective, proactive" style of customer service.

When not working or writing, Tony surprises the children of his urban neighborhood with an ongoing demonstration that vegetables really come from gardens — not from supermarkets. (The neighboring adults are surprised that anything with so many weeds still yields crops, an approach that Tony explains to Mary Frances, his long-suffering wife of a dozen years, as really "low-intensity agriculture at its finest.") After confessing this sin and others generally related to his persistently Rabelaisian sense of humor, the chastened author is to be found on Sundays alongside his wife in St. Paul's Cathedral, the inner city Catholic church of Worcester, where together they serve as lectors.

# Acknowledgments

This work would never have been completed without:

*Joel Villa,* Coordinator of Audiovisuals, College of the Holy Cross. In a world where everyone else is focused on the bottom line, he's got the big picture. Every graph in this book has been provided by him.

*William Littlefield* and *Carolyn Mills,* past and present Science Library Assistants at Holy Cross. When my mind is somewhere in outer space (usually at QB991) they manage the rest of "Q" through "T."

*My student and faculty customers.* Their tuitions and grant overheads support me and my library. Waiting on them in my library has taught me just about everything I know. I thank God that they are so inquisitive and so smart, and that I shall never run out of new things to learn.

*Dr. Eugene Garfield,* President of the Institute for Scientific Information. Not only has he allowed me a certain systematic use of his copyrighted materials in this instance (see the Introduction for a more technical acknowledgment with appropriate "legalese"), but he has provided us all with the most provocative and important analyses of scientific journals of the twentieth century.

*Elizabeth Myers,* Copy Editor, for tremendously improving my poor prose.

*Peg Marr,* Proofreader, for keeping typos to the lowest number humanly possible.

# Index to Journal Titles

# Introduction:
# A Working Guide
# for Working Librarians
# from a Working Librarian

This author was not born an expert on scientific specialties, nor were most librarians. Nor, frankly, did most of us come by this knowledge in college or library school. Yet, the overwhelming number of expensive science journals that all of us buy are highly specialized. Librarians like this author have little hope of personally comprehending all the specialized content of these journals, rating that content, and then making our own selection decisions. Yet the author feels that unless we somehow get a handle on why scientists pursue given specialties and find certain specialty journals more attractive than others, our role in journal selection will be reduced to that of waitresses taking faculty orders. That's a risky practice, since, in the journal restaurant, the waitress is always stuck with the tab. Given the money involved, and given our own internal need for some meaningful involvement in the selection process, we must make some sense of the world of scientific specialty journals.

*How can this be done?* The author has tried to figure out three things that might help him in this dilemma:

- First, how did given scientific specialties get started?
- Second, which journals have sprung up to support those specialties?
- Third, which of those specialty journals seem to be good for both the faculty and the library?

The author's approach is based largely on the biography of science for several reasons, not the least of which is that he could never pass the calculus necessary to directly master most advanced scientific topics. Moreover, he has become fascinated by the people throughout history who actually could do the calculus and advanced science. These people were characterized by the urge to open up new side trails off the well-beaten paths of existing science. Their urges seem to offer a better explanation of specialty journal proliferation than the "publisher's conspiracy" theory espoused by many of the author's colleagues with apparently deeper insights into the scholarly press. At the very least, a look at their lives gives this author a feel for what their spiritual successors want by way of specialized journals today.

The author's knowledge of scientific biography is scarcely the result of years of direct personal examination of primary sources during leaves or sabbaticals. (Like many librarians at small private colleges, he has absolutely no faculty status nor institutional directives or incentives to publish.) Rather, it has come to him indirectly, largely through seventeen years of waiting on customers, 40 hours a week, in his library. It is an admittedly incomplete, sporadically acquired inventory. It contains many faculty anecdotes, traces of book contents scanned quickly for relevance to a student's needs, recollections of scientific obituaries in the press, and episodes from public television. Inevitably, some apocrypha has crept in. This author apologizes for any that misleads the reader. Yet this author has become convinced that the more enduring scientific myths have power in the actual shaping of specialty research traditions.

Nonetheless, the author has repeatedly consulted certain standard references over the years that have substantially formed his views. He has no desire to claim as his own the greater originality, diligence of detail, and level of sophistication that these works and their authors display. The author wishes to acknowledge wholeheartedly his dependence on the following sources. Any errors in the recounting or interpreting of their contents are likely the results of the author's inadequacies rather than a lack of clarity on the part of these sources. Likewise, any remarks or characterizations in this text that might be controversial are the author's own, as are all the specialty journal recommendations. Moreover, the notion of an anecdotal history of scientists for the sake of understanding specialty journals is, for better or for worse, entirely the author's. Nonetheless, if it were possible to put giant footnotes in this book, they would include:

Isaac Asimov. *Asimov's Biographical Encyclopedia of Science and Technology*. Garden City, NY: Doubleday, 1972.

Allen G. Debus, editor. *World Who's Who in Science*. Chicago: Marquis Who's Who, Inc., 1968.

Laura Fermi. *Atoms in the Family*. Chicago: University of Chicago Press, 1954.

George Gamow. *Thirty Years that Shook Physics*. Garden City, NY: 1966.

Douglas M. Considine and Glenn D. Considine, editors. *Van Nostrand's Scientific Encyclopedia*. 6th edition. New York: Van Nostrand Reinhold, 1983.

Charles Coulston Gillespie, editor in chief. *Dictionary of Scientific Biography*. New York: Charles Scribner's Sons, 1971.

Dean Hollister, director. *American Men and Women of Science*. 17th edition. New York: R.R. Bowker, 1989.

Aaron John Ihde. *The Development of Modern Chemistry*. New York: Harper and Row, 1964.

Robert A. Meyers, editor. *Encyclopedia of Physical Science and Technology*. Orlando, FL: Academic Press, 1987.

Ellis Mount and Barbara A. List. *Milestones in Science and Technology*. Phoenix, AZ: Oryx Press, 1987.

Sybil P. Parker, editor in chief. *McGraw-Hill Encyclopedia of Science and Technology*. NY: McGraw-Hill, 1987.

Sybil P. Parker, editor in chief. *McGraw-Hill Modern Scientists and Engineers*. New York: McGraw-Hill, 1980.

Bernard S. Schlessinger and June H. Schlessinger, editors. *The Who's Who of Nobel Prize Winners*. Phoenix, AZ: Oryx Press, 1986.

Knut Schmidt-Nielsen. *Animal Physiology*. New York: Prentice Hall, 1960.

John Walton, Paul B. Beeson, and Philip Rhodes, editors. *The Oxford Companion to Medicine*. Oxford: Oxford University Press, 1986.

James D. Watson and John Tooze. *The DNA Story*. San Francisco: W.H. Freeman, 1981.

Harriet Zuckerman. *The Scientific Elite: Nobel Laureates in the United States*. New York: Free Press, 1977.

## WHAT MAKES A SPECIALTY JOURNAL ATTRACTIVE TO A SCIENTIST?

After the biography of a family of specialties has been discussed, the place and relative merits of the journals that sprang from that history will be discussed from two perspectives: scientists with manuscripts and limited reading time, and librarians with limited budgets and shelf space. To supplement the prose, the odd-numbered figures throughout this text report three factors that appear to affect the perceptions of U.S. scientists concerning journal desirability.

*First, the percentage of recent papers from U.S. labs.*[1] Americans concentrate the bulk of their manuscript submissions and reading time on journals dominated by U.S. authors.

*Second, the percentage of papers from scientifically competitive Western Europe, Canada, and Japan.*[2] American scientists are smart enough to know that these countries have journals that not only make for informative reading but for good alternative manuscript placements.

*Third, "relative impact factor."*[3] "Impact factor" might be roughly defined as the total number of recent citations to a journal divided by the total number of its recent papers. It is one statistic among many reported in the *Journal Citation Reports* section of *Science Citation Index* annuals. "Relative impact factor" is this author's own modification of the reported impact factor. It calibrates the leader in any set of impact factors from closely comparable specialty journals at 100, with those of lesser impact set as a percentage of that leader. This author believes, along with the publishers of citation data at the Institute for Scientific Information, that no single method of rating a journal is sufficient to indicate its quality. However, the author feels that "impact factors" are the best form of quality indicators among the many types of citation data. Many capable, PhD'd bibliometricians and librarians disagree, but the most distinguished scientists, voting with their manuscripts, seem to concur with the author. The systematic use and mention of impact factors from the most recently published *Journal Citation Reports*, and certain other terms, data, and information published by the ISI, have been allowed by special permission of Dr. Eugene Garfield, President of ISI. The Institute for Scientific Information, *Science Citation Index*, *Journal Citation Reports*, *Current Contents* and other ISI names or products used in the preparation of this text, or mentioned within, are copyrighted and represent exclusive trademarks and servicemarks. Systematic use in other publications by others without express permission is forbidden.

As practice, let the reader look at Figure 1. Three imaginary, but scarcely atypical, specialty journals are compared: the *American Bulletin*, the *Eurojournal*, and the *International Proceedings*. The *American* has 60% U.S. authorship, the *Eurojournal* has 20%, the *International* has 10%. The manuscript share from other major scientific powers is 20% in the American, 60% in the *European*, and

FIGURE 1

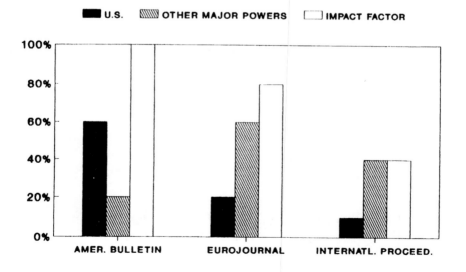

## Contributors and Relative Impact
## Fictional Journals

40% in the *International*. A few things should be noticed immediately. First, rarely do these manuscript market shares add up to 100% for any given journal. Twenty percent of both the *American* and *Eurojournal* titles are "missing," as are 50% of the *International* titles. This share is often for papers from less scientifically advanced countries, or from some countries that are rather isolated, even if their smaller share of papers is relatively advanced.

It is important to note that the author does not consider papers from Third World countries or from the former Soviet Bloc to be inherently "bad." Rather, the author's experience suggests that the best papers from most of those countries are sent by their scientists to American or to major European journals first, and only secondarily to journals that are primarily composed of Third World manuscripts. Conversely, this author does not regard American or even Western European papers as automatically better than Third World papers. Some of the most highly regarded U.S. journals regularly reject American manuscripts in favor of foreign works for up to 40% of their total content. Indeed, the genius of American science is that so much of our talent is imported. (The author, while claiming no special talent, was a refugee himself.[4]) Nonetheless, in a world of finite cash and shelf space limitations, the librarian might make a selection that favored the *American Bulletin* first, the *Eurojournal* second, and the *International Proceedings* last. It should be kept in mind that even last place is generally honorable because there are so many other journals that did not even qualify in this author's judgement for the final comparison set. This book treats fewer than one thousand journals to one degree or another, certainly less than 5% of the world's extant scientific periodicals. In any case, a look at the impact factors suggests that the librarian's plan of incremental selections in this group is confirmed by the independent citing behavior of scientists in this specialty.

## WHAT MAKES A SPECIALTY JOURNAL ATTRACTIVE TO A LIBRARIAN?

The even-numbered figures throughout this text indicate three factors that give the librarian some sense of the relative value of closely matched journals in a scientific specialty.

*First, relative number of papers per year.* This factor indicates how many articles, both full-length, and brief communications or "letters" of original research, are carried by the journal annually as compared to its competitors in that same year. The journal with the most is calibrated at 100, those with less are set as percentages of that leader. (Journals that are primarily devoted to lengthy review articles or are solely devoted to paragraph-sized summaries of convention presentations are *not* included in these comparisons, but may be discussed elsewhere in this text.)

*Second, relative annual subscription price.*[5] Subscription prices were taken for matching years from a variety of sources. However, the same method of price determination is used consistently within any given set of matched journals. The highest subscription price is calibrated at 100, and those that cost less are set as percentages of that cost leader.

*Third, relative subscription base among selected U.S. libraries.* Circulation figures for many scholarly journals are closely guarded secrets. The author has attempted to approximate the relative rate of these journals being held by OCLC member libraries through a tally of reporting libraries that show up on "dha" screens in response to an ISSN search. This method is certainly imperfect since not all subscribing libraries are OCLC members, nor are all OCLC libraries that report some holdings necessarily current subscribers. Yet, as long as the method is consistently applied, a rough estimate is possible.

Figure 2 tells the tale from a librarian's perspective. The *American Bulletin* has the most papers, the *Eurojournal* has 75% as many, and the *International Proceedings*, about 60%. The most expensive journal is the *International*, the next most expensive is the *Eurojournal* at 75% of that cost, and the least expensive is the *American* at 40%. A library selector who had opted for the *American*, followed by the *Eurojournal*, and, if funds allowed, the *International*, would be in concurrence with most of their colleagues among the sampled libraries. It is important to note that this author has not based his recommendations within this text solely on the apparent popularity of a title. As with any library situation, local circumstances may suggest a different course. It should be remembered that both the relative newness of a journal and its price can

FIGURE 2

## Comparison of Number of Papers, Costs, and U.S. Market Penetration

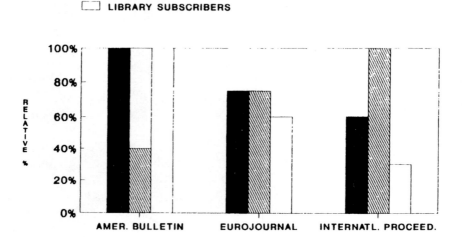

militate against a good showing for a better journal in the "library subscribers" portion of any figure in this book.

The author realizes that each of these six data points are subject to change over time. (Indeed, virtually all of this book was written during 1990.) However, this author feels that any working librarian can determine those changes readily for himself or herself—time well spent in this era of multithousand-dollar journals. It is this author's experience that while individual figures within any matched set of journals will certainly change, these changes do not occur within a vacuum. Neither the publishers nor the editors of these journals are indifferent to the moves of competitors. They monitor what the competition is doing and adjust accordingly. Journals get better not so much through altruism as through the relentless contest for survivability and leadership.

The author attempts to improve through acknowledgment of his mistakes, acceptance of advice from the many colleagues who are more capable than he, and through paying closer attention to his customers. Criticisms of this author and of his work will be greatly appreciated. Send them to this author at the Science Library, Swords Hall 100, College of the Holy Cross, 1 College Street, Worcester, MA 01610-2395.

## NOTES

1. Journals in a set of matching specialties, appearing during matching periods, had 100 articles — or six month's output, whichever came first — examined. In cases of multiple authorship with authors from differing countries, the affiliation of the apparent senior author was decisive. As a practical matter this meant assignment of the paper to U.S. or major Western Powers authors in most cases.

2. For the purposes of this book, scientifically competitive Europe is defined as the U.K., France, the Benelux countries, Switzerland, the former *Bundesrepublik* portion of Germany, and Scandinavia. This necessarily selective judgement was assisted by the ongoing analyses of Hungarian bibliometricians A. Schubert, W. Glanzel, and T. Braun, which appear on a regular basis in *Scientometrics*.

3. Eugene Garfield, Editor in Chief. *SCI Journal Citation Reports, 1988 Annual*, vols. 19 and 20. Philadelphia: Institute for Scientific Information.

4. This author's birth was more remarkable for its characteristically "American" nature than for anything else. His Lithuanian father, his Latvian-born German mother, and his two brothers, born in refugee camps near Denmark, came over to the U.S. in late 1950. The author was born in early 1951 in Worcester City Hospital, where his mother was assisted in the delivery by an intern, himself freshly arrived from Punjab, India.

5. Prices were taken variously from those published within the journal issues, from jobber's price lists, publisher's catalogs, and from standard periodical reference works. Within each matched set, the method of subscription rate determination was consistent.

# Chapter 1

# Analytical Chemistry and Its Journals

## BACKGROUND

Analytical chemistry is both one of the oldest and most modern of the major chemical specialties. Practical men and women as well as scientists have long been concerned with "What's in it? What's this mixture made of? What impurities are still there?" Historically, analytical chemists have sought first to detect the presence of an element or compound, and then to measure its amount, usually by separating it out. Today, journals that report such analyses are taken at over 3,000 libraries in the U.S. alone.

### A Good Start, but a Perilous Adolescence, for Analytical Chemistry in General Chemistry Journals

Articles dealing with analytical chemistry were initially welcomed into early chemistry journals, for they accurately reflected the utilitarian nature of early science. Research topics were chosen for their practicality. The results and discussion of analytical papers tended to be specific and limited in their relevance for other types of chemistry. Analytical chemists tended to solve problems, not propose new laws of chemistry.

In a way, analytical chemistry was spoiled by its early successes. Carl Fresenius, a German university professor of analytical chemistry in the mid-1800s, went from triumph to triumph during a very long career in the service of agriculture, mining, and manufacturing. He invented dozens of instruments along the way, effectively creating what scientists today call "wet" (flask and test tube) analytical chemistry. His many students tended to dutifully emulate Fresenius and improved on his methods and tools only incrementally, not conceptually. Few attempted genuine innovations in their approach to analysis. It was not that these young chemists were lazy. They often went to heroic efforts to get a refinement of a standard procedure. But they often felt they succeeded when they had solved the immediate problem of identification and quantity measurement, and stopped before they had gained insight into the underlying reason for their success. Moreover, in that era of academic and political rivalries, research and training in analytical chemistry had a strong Teutonic flavor that was distasteful to those who envied the success of the Germans, but thought them uncreative. This "German" science was regarded by some as something to be used, not admired; much in the way that some Americans find cars from Japanese manufacturers to be of high quality, but the Japanese auto worker's devotion to his employer alien to a free-thinking lifestyle. To further the matter of resentments, Dutch, German, Austrian, and Swiss manufacturers held a dominant position in chemical instrumentation through their patents for the finest test chemicals, the best balance scales, and other analytical apparatus.

The control of the supply and price of this technology was important because analytical chemists of the era preceding the world wars relied largely on precision measurement of obvious characteristics like melting point, density, or the appearance of particles under a microscope. Many of their testing methods were based on isolating the target substance by painstakingly dissolving the sample and recrystallizing it, over and over. These tortuous procedures often caused the destruction of the major portion of the sample along the way. Indeed, the notion of a "sample" was very different in the era of Fresenius. To get an idea of the quantity of an ingredient or contaminant, very large "samples" would often be required. Even as late as 1911, the Nobel Prize-winning isolation of radium by Marie Curie was classic in this regard: a half ton of ore yielding less than a single gram of radium. After a while, results obtained by analysts reached the limits of existing "wet chemical" technology. Few clear-cut breakthroughs seemed to develop. Electrochemistry, based on analyzing dissolved materials in solutions using varying voltages and electrodes, was probably the most advanced. It, too, was largely a German province under a another pioneer named Kohlrausch. But the typical electrochemical set up of that time resembled today's car battery with the addition of some meters. It was not until 1959 that a Nobel Prize was awarded to an electrochemist, Jaroslav Heyrovsky, a Czech, and some critics thought that was merely acknowledging a piece of history rather than a recent breakthrough.

The journal literature of wet analytical chemistry matched the laboratory experience. It, too, tended to increase in bulk far more quickly than it did in fashionability, depth, or sophistication. For example, spot tests — "a little iodine into a milky solution yields a brown color if there is starch present" — were recorded and collected by analytical chemists much as a cook collected recipes. Fritz Feigl, an Austrian chemist most active in the 1930s, recorded literally over a thousand such spot tests in a handbook whose current edition runs almost fourteen hundred pages. Indeed most current at-home pregnancy test kits involve one or more of its procedures.

These many approaches to chemical analysis certainly worked well enough for the situations faced at the time of their devising, but there was a lack of a central, organizing, intellectual theme or rallying point for academic scientists. By the 1930s many prestigious eastern universities in the U.S. downgraded analytical chemistry as a specialty worthy of PhD training. It was thought to be a dead-end technician's job. Papers in analytical chemistry became subtly unwelcome in several major chemistry journals that were supposed to cover all fields of chemistry.

## Out on Their Own, Analytical Chemists and Their Journals Surprise the Critics by Growing Strong

But four major factors sustained analytical chemistry and furthered the founding of journals specifically devoted to analytical chemistry:

1. The continued necessity for chemical analysts in the "real" (nonacademic) world
2. The pluralism of American higher education (not every school was, or wanted to be, the Ivy League)
3. A fortuitous concentration of analytical research on two main

approaches to analysis that had a foundation in more basic sciences that did have academic panache

4. The rise of microelectronics and computerization in analytical instrumentation, enabling combinations of methods and allowing for sophisticated interpretation of results with much less destruction of the sample.

Not all of America's strong universities were swayed by intellectual fashionability to the point of eliminating analytical chemistry. It was politically impossible to abandon a field that contributed so vitally to local economies. Instead, particularly in the agricultural midwest, criticisms of the cookbook method of teaching analytical chemistry tended to foster curricular improvement. Prospective analytical specialists from reformed programs got an educationally sound sequence of courses in other scientific disciplines. Young analysts went from being memorizers of old lore to monitors of current scientific literature. Their use of journals increased.

The growing demand for quality control in industry also meant a demand for more analytical chemists. In the newly developing age of the continuously running assembly line, engineering schools developed instruments that shifted the emphasis from stopping production and examining huge quantities of product to the continuous monitoring of small amounts of the material as it was being processed. This was in great measure possible because of the familiarity of engineers with the earliest advances in electronics and the long-standing engineering school expertise in devising controls for flow, pressure, and heat. Engineers simply developed electronic ways of detecting changes in a substance that substituted in whole or in part for traditional wet chemical tests. The analytical lab in industry quickly changed from a jungle of glassware, pulverizers, and balances to an automated array of consoles with periodic read-outs.

A major impetus towards continued work on small samples came from medicine. Blood, for example, could not be extracted from the patient in unlimited quantities just to make life easier for the chemical analyst. Analytical chemists in the service of hospitals or in the pharmaceutical industry learned to monitor small changes in small samples. Indeed, Fritz Pregl, an Austrian surgeon, gave up operating on patients in order to perfect what he called the "Microscale" clinical analysis of bile from the liver, and won a Nobel Prize for it.

Each of the constituencies—agricultural, industrial, and medical—that sustained analytical applications gave rise to journals of applied analytical chemistry that remain important today.

## *Spectroscopy and Chromatography as Rallying Points for Analysts and Topics for New Journals*

The intellectual cachet of analytical chemistry further improved when the many disparate methods traditionally used by analysts tended to coalesce into two broad main conceptual approaches. While advances in wet analytic chemistry and electroanalytical chemistry have never really stopped, emphasis for the bulk of the twentieth century has centered on spectroscopy and chromatography.

Spectroscopy is based on studying the wavelengths of the energy emitted or absorbed by matter. An emphasis on spectroscopy was not only very fruitful for new techniques in analytical chemistry, but it gave the field strong ties to the highly esteemed optical physics community, and a theoretical foundation. Indeed, it was a partnership between the developer of a famous piece of laboratory equipment, Robert Bunsen of Bunsen burner fame, the last great wet analytical chemist, and Gustav Kirchhoff, a physicist with a strong interest in identifying the elements within the sun, that led to the invention of the spectroscope. Spectroscopic studies were highly amenable to mathematical analysis and theoretically intriguing. They brought analytical chemistry data to a circle that included Max Planck, Kirchhoff's student and a Nobel Prize winner as founder of quantum mechanics; and to the young Einstein, who won the Nobel Prize for the photoelectric effect, a quantum mechanical phenomenon on which much spectroscopy depended, long before "relativity" became a household word.

Analysts had long realized that the underlying principles governing old techniques such as pyrolysis (analysis by flame) and colorimetry (analysis of the concentration of an ingredient based on color comparisons) were really just individual examples of the many chemical characterizations possible using light waves. But their system of argumentation in the literature and exposition in the class-room had not given much evidence of any systematic thinking. With the incorporation of modern physical theory this changed for the better. The final conceptual breakthrough that energized the an-alytical community came in response to the following question: "Why are we using only visible light waves for spectroscopic anal-ysis when physicists have already done so much preliminary work for us in other wavelengths?" Analytical leaders began to exhort students and colleagues that vast numbers of pure, unapplied stud-ies — usually performed by physicists and involving electromagnetic waves that we cannot see — had enormous potential for chemical analysts who had a sufficient grasp of the new quantum wave theo-ries. Over the years analytical specializations and the journals that would cover them sprang up to fill the entire electromagnetic spec-trum: microwaves, infrared and ultraviolet light, gamma and X-rays, etc. Some of the methods that arose were more useful for detecting individual elements, giving rise to journals favoring atomic spectroscopy; other procedures favored more complicated compounds, giving rise to journals of molecular spectroscopy. In many cases physicists and physical chemists still contribute to these journals. In some cases they still dominate them.

Chromatography provided the second rallying point for modern analytical chemistry. In its simplest form, chromatography is an analytic method by which mixtures within a fluid are sorted out by attrition as they flow through a generally thicker medium. In an everyday analogy, chromatography is like a family footrace through an obstacle course. If such a race followed the rules of chromatog-raphy, the smallest, most highly mobile children would finish first, the middle-sized parents would finish next, and then a corpulent uncle such as the author would finish last.

If spectroscopy cemented the ties of analytical chemistry to the physical sciences, chromatography extended its ability to serve the life sciences, particularly the fashionable world of biochemistry. The "chroma" in chromatography, refers to color, much as it does in trade names like "Kodachrome™" brand color film. This early emphasis on color came from the experiences of many analysts as they broke down the green and orange juices of plants into bands of differing shades. These bands of color would become quite distinct as the test solution traveled through glass columns packed with uniform fibers or granules. Before chromatography, separations of the differently colored compounds were very laborious, and it is ironic humor that much of the credit for this wonderfully labor-saving approach is largely due to two scientists named Tswett and Strain ("sweat and strain")!

Sometimes the procedures were disarmingly simple. Often strips of filter paper, hanging with tips dipped in test solutions, gave interesting results, in a process called, sensibly enough, paper chromatography. Over the years, more and more variations of liquids through gels, or thin liquids through thick liquids were developed into new chromatographic tools. With time it became apparent that electronic means of spreading separations or detecting individual bands of separated material were often more exact than visual inspection. Electronic means are particularly necessary when gas is used as a force driving the test material through specially treated tubes, in what is termed gas chromatography. Today, combinations of these tools are quite common, and Nobel Prizes in medicine involving work in chromatography have been frequent since 1952. In that year Archer Martin received the award for his work with a liquid-through-liquid type of chromatography. He would later go on to greatly improve both paper and gas chromatography. Today there is a large number of journals covering both general and specific forms of chromatography.

## Ironically, as the Instrumentation Shrank in Size, Analytical Chemistry Grew "Important Enough" To Be Welcome Again in General Chemistry Journals

Today, at least partly because of the revolution in microelectronics, analytical instruments are more compact and more reliable than ever. The largest part of many instruments is the TV screen on which the results are displayed with graphs and numbers. Printers connected to instrumentation today give publication quality graphs that are the results of dozens of samples rechecked numerous times within the course of an hour. Samples may sometimes be processed both chromatographically and spectroscopically with a single injection into a test chamber. The sample sizes needed today are often tiny, and the instruments are sometimes not destructive of even those small amounts. So many other specialties need analytical tools, and the uses for analytical chemistry have increased so rapidly that the analytical chemist is often the best-equipped, best-funded, and most consulted member of the chemistry staff. Several journals of analytical chemistry are among the less expensive in chemistry, despite their bulk and their exceptional amount of illustration. This is due to the tremendous amount of glossy advertising found in the journals. Makers of instruments and supplies know that these are the outlets for their best customers.

Nonetheless, in light of historical circumstances, it is heartening to note that in 1982, an unabashed electroanalytical chemist, Allen J. Bard of the University of Texas, assumed the job of editor in chief, of one of the world's most cited general chemistry journals, the *Journal of the American Chemical Society*. Interestingly, it is not clear, even in 1990, that many analytical chemists are willing to leave the thriving family of journals long devoted to their special needs to accept the invitation this development implies.

## SORTING OUT THE JOURNALS

### Journals of General Analytical Chemistry

There are over twenty journals purporting to cover all aspects of analytical chemistry. Most are characterized by some variant of the word "analytical" in the title. *Analytica* is a neo-Latinism. *Analusis* is not a misspelling. It is the Greek word for analysis, and ironically designates the French journal for analytical chemistry, most of whose papers are now in English! Virtually every large country has at least one analytical journal, and most of today's international journals have roots in what was originally a national journal. While not-for-profit societies devoted to chemistry are often the sponsors of these journals, the for-profit sector has expanded very rapidly in this field because of its lucrative subscription base and possibilities for income through advertising.

## A Journal in a Class by Itself:
**Analytical Chemistry**

The leading journal by any measure is *Analytical Chemistry*, from the American Chemical Society. It had somewhat humble origins, existing from 1929 to 1946 as a supplement to *Industrial and Engineering Chemistry*. It is now twice as cited as its nearest leading competitor, and attracts the best American work. In an age of complaints about costly scientific journals, it is a bargain with over three thousand densely illustrated pages annually for less than ten cents a page. In each even-numbered year, the journal provides a special issue of very extensive review essays covering progress in the fundamental methods within analytical chemistry. Each odd-numbered year features similar works on applications of analytical chemistry in industry, agriculture, medicine, etc. Free annual laboratory buying guides are another extremely useful feature.

### Other Leading Journals

Library selection becomes more complicated after one takes *Analytical Chemistry*, since the field is rich in other worthwhile titles at the general analytical level. Figures 1 and 2 compare some of the more prominent titles. The *Analyst* comes from the Royal Society of Chemistry, and has a strong heritage (1876) of not only concentrating British work, but that of many of the Commonwealth countries. *Fresenius Zeitschrift fuer Analytische Chemie* is at least as historically important. Founded in 1862 by its famous namesake, it has the overwhelming support of chemists in Germany, Austria, and Switzerland. Now handled by the strong Springer publishing empire on behalf of the German Chemical Society, many of its papers are in English.

While both of these essentially national journals accept good papers from any land, three journals have been decidedly international from their post World War II inceptions. *Analytica Chimica Acta* comes from Elsevier. It vies with *Fresenius* for the largest number of papers, and succeeds in attracting many from the better known scientific centers. *Analytical Letters* takes the fewest papers, and is the youngest (1968) among the journals. Based in New York, it stresses quick acceptance or rejection as an attractant. As with most

FIGURE 1

## Contributors and Relative Impact
## Leading Analytical Journals

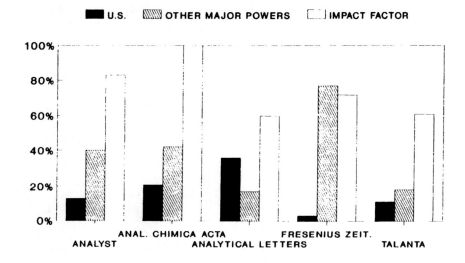

Marcel Dekker titles, it uses the camera-ready-copy method of page production to help deliver on its claims of speed. The third international title, *Talanta*, is a Pergamon product. Two of its more distinguishing features are special symposia issues focusing on given topics within analytical chemistry and surveys of the state of analytical chemistry in given countries.

While the very largest working analytical laboratories and graduate programs will want all of these general analytical titles, and perhaps several others, small programs in U.S. undergraduate colleges are faced with tough choices. The *Analyst* is undoubtedly the least expensive per paper, and has many papers from scientifically advanced countries, but has only about 13% U.S. papers. While *Fresenius* has the highest proportion of papers from the developed West, most are from its historic Central European constituency. It has the lowest level of contributions from the U.S. — less than 5%.

FIGURE 2

## Comparison of Number of Papers, Costs, and U.S. Market Penetration

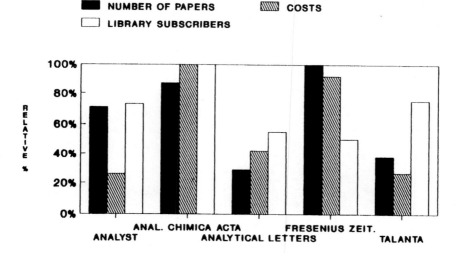

*Talanta* has some price advantages, but many of its papers are from scientifically less competitive countries, a factor reflected in its impact factor. Its U.S. share is also modest.

*Analytica Chimica Acta*, on the other hand, is highly competitive in terms of its pool of contributors, managing to include some good British and German papers despite other strong journals from those countries. Virtually all of its papers are in English, another advantage over *Fresenius*. It has a very large number of papers overall and one of the best impact factors. Despite its price, it might be favored in that it is the foreign outlet most commonly used by American authors in this field.

If funds allow, *Analytical Letters* is a good third choice. It has the highest proportion of American authors, particularly from the industrial sector. When this factor is coupled with its speed and rather moderate price, it may offset the journal's more modest gross number of papers and their modest impact factor.

A field like analytical chemistry that produces so many scattered articles cries out for journals of evaluative essays that summarize and integrate large numbers of those papers. While the scheduled reviews in fixed categories within *Analytical Chemistry* serve that function well, some find the more varied and more frequent assortment of extensive reviews within *Critical Reviews in Analytical Chemistry (CRC* — Lewis Press) a desirable complement.

The brief, invited minireview is the staple of two journals. *American Laboratory* (1968) from International Scientific Communications, tends to emphasize U.S. developments. Elsevier's *TRAC: Trends in Analytical Chemistry* (1981) is an important European counterpart. The latter title provides an option of sending libraries the whole year, completely bound, at the end of the year, much like a Wilson index.

### Journals of Spectroscopy

While there is also a clear first choice in this field, that distinction comes with a few qualifications. *Applied Spectroscopy* has long been the principal outlet for American analytical chemists who choose a spectroscopy journal. It shares many strong features and price advantages with *Analytical Chemistry*. However, as with all spectroscopy journals, some of its papers will be of greater interest to physicists and optical engineers, a development explained by the academic roots of spectroscopy. Moreover, *Applied Spectroscopy* is not clearly the leader in impact factor worldwide. Some of the quasi-specialized spectroscopy journals from the for-profit sector and from Europe share that ranking.

Specialization in spectroscopy journals operates on at least three levels. This discussion concerns the leaders on the two levels of specialization of the greatest concern to most chemistry libraries. On one level are two truly general spectroscopy titles: *Applied Spectroscopy* and *Spectroscopy Letters*. On the next level are those journals which specialize more broadly in either atomic or molecular spectroscopy. This modest level of specialization involves a variety of wavelengths and many different spectroscopic tools. Because of their reasonably broad scope, most working laboratories and academic programs will take one or more of these atomic or molecular titles. Spectroscopy or optics titles that are further sub-specialized on a third level, by either a specific range of wave-lengths (e.g., *International Journal of Infrared and Millimeter Waves*, Plenum) or by a single analytical tool (*Journal of Raman Spectroscopy*, Wiley) are not discussed here. Their value may be quite great to physics programs and to given chemistry programs, but this is highly dependent on the particular types of instrumentation available there.

Figures 3 and 4 show the journals that best meet our wider interest criteria. *Spectrochimica Acta*, a Pergamon title, is actually two sibling journals derived from a parent journal established in 1939. Its section "A" has long handled some of Europe's best papers in molecular spectroscopy. This section competes most closely with the *Journal of Molecular Spectroscopy*, an American entry from

FIGURE 3

## Contributors and Relative Impact
## Leading Spectroscopy Journals

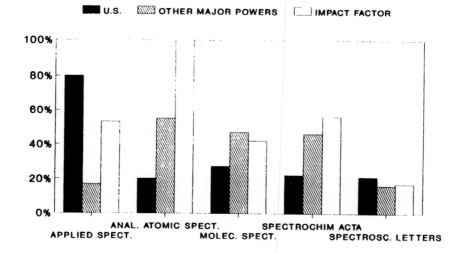

FIGURE 4

## Comparison of Number of Papers, Costs, and U.S. Market Penetration

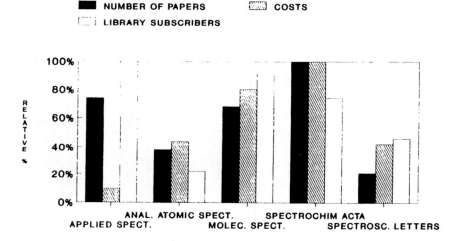

Academic Press with a decided physics flavor. *Spectrochimica Acta B* deals with atomic spectroscopy, and as such competes most closely with the Royal Society of Chemistry's *Journal of Analytical Atomic Spectrometry*. Finally, *Spectroscopy Letters* is a Marcel Dekker title with more scope and speed than any of these more specialized titles, but with many fewer papers. All of these journals share a rather strong level of participation by American contributors.

If the library serves a clientele with a heavy atomic preference, the team of the *Journal of Analytical Atomic Spectrometry* and *Spectrochimica Acta B* is best. The opposite matchup, the *Journal of Molecular Spectroscopy* and *Spectrochimica Acta A*, is better for a molecular emphasis. If balance between atomic and molecular interests is required and funds do not allow for more than two titles, *Spectrochimica Acta A and B* together probably have greater advantages. They have a seniority of high reputation and a more pronounced analytical chemistry flavor.

This leaves us with *Spectroscopy Letters*. The rather modest impact factor, fewer papers, and greater Third World authorship participation of *Spectroscopy Letters* makes this journal somewhat less desirable, despite its characteristic Dekker advantages in manuscript turnaround time. The *Canadian Journal of Spectroscopy* (Spectroscopy Society of Canada), not shown, may, in fact, soon surpass it as a secondary choice, at an even better subscription price.

### Journals of Chromatography

The selection of journals of chromatography is somewhat complicated by the lack of a world leading title from a U.S. society, and some very strong European entries that could lay claim to that distinction. (See Figures 5 and 6.)

The oldest, largest, and one of the more cited titles from Europe is Elsevier's *Journal of Chromatography*. With over one thousand

FIGURE 5

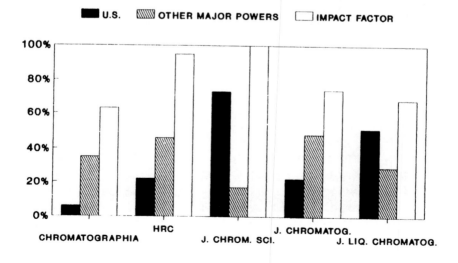

## Contributors and Relative Impact
## Leading Chromatography Journals

■ U.S.   ▨ OTHER MAJOR POWERS   ☐ IMPACT FACTOR

FIGURE 6

## Comparison of Number of Papers, Costs, and U.S. Market Penetration

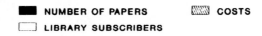

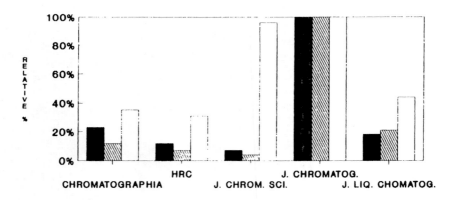

papers a year, it represents an indispensable title for all of the best research centers in the scientific world. But with an overall bill in the multiple thousands, it also represents a financial burden for the less wealthy library. Elsevier's per-article price is actually quite competitive, but the library gets and pays for hundreds and hundreds of papers other than what the clientele may wish for as well. Both other European firms and some American ones have mounted challenges to the *Journal of Chromatography* based on offering fewer, but hopefully somehow more select, papers at a greatly reduced individual subscription price.

Of the two most prominent European competitors to the *Journal of Chromatography, Chromatographia* (distributed by Pergamon) offers the second most papers among all chromatography journals and a longer history (1968) even as *HRC: the Journal of High Resolution Chromatography* (Huthig, 1978) offers higher impact factors, and lower costs. An edge might be given to *HRC*, since it attracts a higher share of American papers.

The American contingent is represented by the *Journal of Chromatographic Science* (Preston) and the *Journal of Liquid Chromatography* (Dekker). (The latter title is very much a mainstream title, and does not represent a narrow subspecialty.) The *Journal of Chromatographic Science* comes closest to being the leading American outlet for chromatography with its 73% U.S. authorship, but even so, about twice as many American papers appear in the *Journal of Chromatography*. The *Journal of Chromatographic Science* is, however, the leader in impact factor among all chromatography journals, and is the most reasonably priced. The *Journal of Liquid Chromatography* also has a high U.S. author involvement — 50% — but does not compare as favorably in impact factor or costs. As a Dekker title, however, it is characterized by fast manuscript turnaround time, and an above-average number of papers from working industrial labs.

If, on careful reflection, a library decides that it cannot afford to subscribe to the *Journal of Chromatography*, it will get good global coverage and about half the papers at half the price, if it buys all four competitors. A problem of geographic imbalance occurs if the library takes fewer titles than these.

While the title most likely to be chosen first, the *Journal of Chromatographic Science*, has many good qualities and is very highly U.S.-oriented, this latter strength may require some geographic balancing. The journal that might seem to be a good second choice — the *Journal of Liquid Chromatography* — would lead to mostly American papers and a smattering of foreign papers, with most of either variety being among the field's less-cited. At the very least, for reasons of more exposure to non-U.S. work, it might be wiser to choose a European title. *HRC* is probably a better choice among the two European titles. Not only does it give you much of Europe, it has a demonstrably better record of attracting U.S. papers than *Chromatographia* as well. The *Journal of Liquid Chromatography* remains a good final choice.

Interestingly, chromatography is such a huge field, it now has its own news and practical reviews magazine, *LC-GC* (Liquid Chromatography-Gas Chromatography) from Aster Press. The title is relatively inexpensive and is recommended wherever there are several chromatographers.

### Other Analytical Techniques and Their Specialty Journals

The *Journal of Electroanalytical Chemistry and Interfacial Electrochemistry* is the traditional first choice for analytically inclined electrochemists. Neither the U.S.-based *Journal of the Electrochemical Society* (Electrochemical Society) nor Pergamon's *Electrochimica Acta* is quite as focused on analytical aspects, although both are worthy members of any comprehensive collection. The *Journal of Electroanalytical Chemistry* is, however, another titanic Elsevier title, reminiscent in advantages and disadvantages of its cousin, the *Journal of Chromatography*. *Electroanalysis*, a 1989 entry from VCH, the German chemical publishers, has made a promising start as a competitor with fewer papers at a more easily managed price.

The wet analytic tradition is still alive, but usually in combination with other techniques. *Microchemical Journal* from Academic is probably the first choice of many U.S. researchers. *Mikrochimica Acta*, founded in 1937, in the famous Austrian tradition of microanalysts, is a good second choice with a decidedly European flavor. It is published by the Vienna branch of Springer.

Two instrumental methods of analysis not previously mentioned are nuclear magnetic resonance and mass spectrometry. The first method is closest in function to the other forms of spectroscopy, although technically it depends on the nucleus as opposed to the electron shells, upon which other spectroscopy depends. The second method, despite its "spectro" name, is most often used in conjunction with gas chromatography. Both tools started out in physics and retain much interest for chemists in specialties other than analysis, particularly organic chemistry in the case of nuclear magnetic resonance. Nuclear magnetic resonance (Nobel Prize in Physics for Bloch in 1952) is most prominently reported in Academic's *Journal of Magnetic Resonance*. *Magnetic Resonance in Chemistry* from Wiley is a good journal of second choice. Mass spectrometry (Nobel Prize in Physics for Aston in 1922) is covered in Wiley's *Or-*

*ganic Mass Spectrometry* and Elsevier's *International Journal of Mass Spectrometry and Ion Processes*.

## Journals in Applied Analytical Chemistry: Agriculture and Food Processing

Since its founding in 1915 by what was then called the Association of Official Agricultural Chemists, the *Journal of the Association of Official Analytical Chemists* has been the leading organ of those who analyze and certify foods. The American Chemical Society joined the field with the *Journal of Agricultural and Food Chemistry* in 1953. There is substantial specialization within the field, with journals for every major food group or type of farming. While a title like *Cereal Chemistry* (American Association of Cereal Chemists) is self-evident, the *Journal of American Oil Chemists Society* is not. It deals primarily with vegetable oils and animal fats, not with petroleum products. The group simply refused to continue with the originally proposed title: *The Journal of the American Fat Chemists Society*!

Most of the general professional journals, like the *Journal of Food Science* and *Food Technology* from the Institute of Food Technologists in Chicago, carry analytical papers relating to health concerns. While virtually every major foreign country has journals reporting food chemistry, Blackwell's *Journal of Food Technology* is probably the first choice for multinational coverage in analytical labs that cannot collect individual national titles comprehensively.

## Journals of Applied Analytical Chemistry: Industrial Chemistry and Analytical Instrument Manufacturing

The American Society for Testing and Materials is the leading association for materials analysts, purchasers, and bulk handlers in virtually all major industries. It issues hundreds of irregular bulletins for given products and specifies levels of purity and sampling techniques. Its most general interest title is the *Journal of Testing and Evaluation* (1966). One title useful for industries that involve heat processing or analyzing the products of combustion is the *Journal of Analytical and Applied Pyrolysis*, from Elsevier. Plastics and circuit boards are particularly susceptible to analysis by heat. *Thermochimica Acta* (Elsevier) and the *Journal of Thermal Analysis* (distributed by Wiley) generally involve, in their analysis, temperatures less than the flash point of the material tested.

Among journals more specifically devoted to chromatographic methods, the titles from Dekker stand out. While they do not seem to attract as many citations as journals which have a more academically rigorous tone, they tend to have many highly specific, practical papers, including some from the growing biotechnology industry. In addition to those titles mentioned in earlier sections of this chapter, *Separation Science and Technology*, *Separation and Purification Methods*, and *Solvent Extraction and Ion Exchange*, may be appropriate.

The instrumentation business is extremely dependent on analytical chemistry and vice versa. Dekker's *Analytical Instrumentation* (1968) is a leading title with a chromatographic emphasis. Makers of spectroscopic instruments will want *Applied Optics* and the *Review of Scientific Instruments*, available from the American Institute of Physics. *Chemometrics and Intelligent Laboratory Systems* (Elsevier) is representative of the ever increasing links between automation, miniaturization, and computerization in the analytical lab today.

### Journals of Applied Analytical Chemistry: Medical, Pharmaceutical and Forensic Sciences

As anyone who has been in a hospital in the last twenty years can confirm, an enormous number of "tests" of bodily fluids are constantly being conducted. This is probably the most widespread analytical chemistry of all, and you, as patient, provide the chemical samples. Every hospital and most analytical chemistry collections should have *Clinical Chemistry*, the official journal of the American Association for Clinical Chemistry since 1955. *Clinical Biochemistry* (1967), from the Canadian Society of Clinical Chemists, is an attractively priced companion. *Clinica Chimica Acta* (1956) is a relatively expensive, but characteristically expansive Elsevier entry, with good international coverage.

*Analytical Biochemistry* from Academic is essentially the PhD-program journal of fundamental research for this field. The pharmaceutical literature provides many examples of analysis in its regular journals. Two of the more prominent specialized journals are the *Journal of Analytical Toxicology*, from Preston, publisher of the *Journal of Chromatographic Science*, and Pergamon's *Journal of Pharmaceutical and Biomedical Analysis*. The *Journal of Laboratory and Clinical Medicine* (Mosby for the Central Society for Clinical Research) is the leader among the many pathology journals in terms of publishing MD-authored analytical papers.

A fascinating, if sadly booming, sector of the analytical literature is forensics. It deals with investigating criminal matters: spot testing narcotics, detecting poisons, identifying victims or perpetrators through body fluids or tissues analysis, establishing paternity in matters of support payments, and so on. There are over four thousand subscribers employed in hospital pathology departments and law enforcement settings taking the leading title, *Journal of Forensic Science*, a publication of the American Academy of Forensic Sciences. Good second choices may include the British entry, *Journal of the Forensic Science Society* or Elsevier's *Forensic Science International*.

# Chapter 2

# Inorganic Chemistry and Its Journals

## BACKGROUND

Historically within chemistry, the term "inorganic" referred to compounds that did not contain the element carbon. This special deference to carbon came from the notion during the 1700s and 1800s that the most important biological or "organic" compounds were made solely by nature and largely of carbon. By the late 1800s, the academic field of carbon-containing "organic" compounds became consciously detached from "inorganic" chemistry. With at least 102 elements other than carbon for chemists to study, "inorganic" would seem to be the dominant specialty within chemistry. This is not the case. The field probably accounts for about 10% of the literature today, and would be even smaller if its adherents had not developed a highly flexible and expansive approach to the subjects that properly fall under its domain.

### A Foundation Built on Both Stone and Myth

The primary drive for the knowledge that eventually developed into inorganic chemistry was practical. All of the world's metals were historically derived from mixtures of mostly inorganic elements. Likewise, common household ingredients like table salt and alum for pickling, the clay and chalk used in pottery, glassware, and many dyes were all familiar inorganic materials. The field was further spurred by the development of better bricks, tiles, mortars, and housepaints, materials all substantially composed of inorganic materials. Military needs for sulfur and phosphorous for explosives further increased demand. Early medicines and pest poisons depended on the use of arsenic, mercury, and bismuth.

A secondary drive for what was to become inorganic chemistry came from alchemy. Alchemy was an early marriage of chemistry and folklore with the goal of producing magical results. The principal goal of alchemy involved turning common metals like iron, copper, or tin into gold or other precious metals. While the miraculous transmutation of these metals would never take place, an enormous amount of chemistry evolved concerning the breakdown of these metals and related mineral ores through either heat or dissolving in acids. And although the usage is now changing, inorganic chemistry in France is still frequently termed "chimie minerale." To a lesser degree, the testing of crystals and gems for authenticity and quality set the stage for further advances in inorganic chemistry.

### Inorganic Chemistry Becomes a Matter of Refinement and Gases Come Under Its Domain

While it is completely understandable that the mining and metals industries would have stimulated the need for inorganic chemists in the 1700s, it is surprising that gentlemen philosophers embraced it so fervently. This era saw a new fashion in intellectual circles shifting emphasis from far-away astronomy and abstract mathematics to demonstrations of scientific principles operating right here on earth. Chemistry had a tremendous advantage in that its discoveries could often be made entertainingly obvious to the audiences at increasingly trendy meetings of science academies in major cities. For many member *savants*, however, the emphasis would not lay on the solid, mineral side of inorganic chemistry, but on understanding gases and the nature of combustion.

Combustion as we understand it today generally involves carbon, suggesting that organic chemistry would dominate this field. But earlier interpretations of combustion depended on alchemical notions of "phlogiston" being added or taken away in the material that was burned, rendering it a part of inorganic chemistry. A host of gentleman researchers had tackled this problem and others related to the composition of air, and the transformations of heated solids into liquids and then into gases.

Indeed, in a way that would seem odd now, gases became the

major focus of a scientific rivalry between England and France that paralleled the space race of the mid-twentieth century between the U.S. and the Soviet Union. In England, Cavendish, Black, Hales, and Priestley all produced major gases like nitrous oxide, ammonia, and sulfur dioxide, and even elements like hydrogen. But the most widely recognized success went to France. There, a wealthy noble-man working alone, Lavoisier, was to fundamentally throw out the phlogiston theory while working with heated mercuric oxide. The heated oxide gave him a source of pure oxygen. This feat was also accomplished by his English archrival Priestley, but not as effec-tively explained in the scientific salons of Europe. Lavoisier's work, by contrast, led to a fully worked out, oxygen-based theory of combustion. This triumph, and his early demonstrations of the bonding of gases with other solids or with liquids is held by many to be the start of modern chemistry. (The triumph did not help the noble Lavoisier all that much: he was beheaded during the Reign of Terror during the French Revolution.) For our purposes, it is most significant to note that until well into the nineteenth century, the chemistry of gases was largely within the realm of inorganic chem-istry and once again represented an expansion of the specialty. Even today, the chemistry of the noble gases and fluorine-chlorine fam-ilies of gases are primarily published in journals of inorganic chem-istry.

### New Minerals Yield New Inorganic Substances

In the academies outside of London and Paris, papers describing new substances isolated from solids, typically mineral ores, never went out of fashion. In Scandinavia and Central Europe mining was a major source of the underlying wealth supporting members of scientific academies. More than a dozen new elemental solids in-cluding uranium, cerium, palladium, molybdenum, ruthenium, and scandium, all having mineral origins, were discovered in this era. Ultimately, increasing discoveries in both gases and solids needed an organizing guide, and the unlikely pair of a Quaker schoolmaster and an Imperial Russian bureaucrat-scientist were to provide them.

## Dalton Defines the Elements

John Dalton was in many respects the classic bachelor English schoolteacher of the early 1800s. He lived for his schoolwork and yet was so poorly rewarded he had to seek outside sources of income. He turned to lecturing to the general public, and was surprisingly successful. Despite a serious lack of social connections, he also got along well with the monied classes. His dedication and good will endeared him to seemingly contradictory segments of society. Despite the fact that he became a teacher largely on the old apprentice system, and had completed no university program, he maintained one foot in the increasingly university-educated world of the elite, and another foot in the camp of urban clerical and mercantile workers seeking a little enlightenment.

Dalton's earnestness seemed particularly appreciated in intellectual circles. Scientific academies were finally becoming more serious centers for learned discussion and publishers of scholarly literature. Wealthy dilettantes who were tolerated as benefactor-members had become less numerous and less influential over time, and support from members of the rising middle class increased. Facile demonstrations and scientific entertainments were no longer the style of academy meetings, nor were money or parlor tricks sufficient to maintain one's credentials. A serious program of research was required. Dalton, however, realized that the passing tradition was not without merit. He managed to move the educational entertainment format towards the general public, without abandoning the serious respectability he earned through especially penetrating commentaries written for academy members. So, along the way to serious scientific glory, he became one of the first scientists known by the man on the street.

Dalton was intrigued by the work of Boyle, an English gentleman-scientist of the preceding generation. It was Boyle who first put forth to the English scientific community the concept of the atom. But Boyle did not work out many of the details, and interest in the concept was modest. Even today, Boyle's name is invariably mentioned for his gas law, not for his fitful start on atomic theory. Fortunately for chemistry, Dalton's further development and proselytization of definitions for the "atom" and "element" took firmer hold in England. Dalton emphasized constant, reliable, and virtu-ally immutable properties as qualifications for real elements. He essentially gave chemists benchmarks at which to aim in reducing substances to their simplest, purest form. His fame and his atom and pure element concepts eventually spread abroad, and, indeed, a kind of competitive scientific nationalism soon crept into the dis-covery of new elements using Dalton's criteria. To a large degree, this effort replaced the war of the gases.

## Putting Elements in Their Place: Mendeleev and the Periodic Table

One of the principal concerns of scientists nonetheless remained distinguishing between what was genuinely a new discovery and what was merely a new source for an element already discovered earlier in some other type of ore or mixture of gases. Adding to this confusion was that certain clearly distinct elements shared similar properties and ratios of bonding with other elements, a problem even Dalton had tackled with only partial success. Schemes of clas-sifying the elements to account simultaneously for both uniqueness or similarities met with limited acceptance and success.

It was left to Mendeleev, a comfortably-raised, governmentally well-supported, Russian educator of the next generation, to make that breakthrough. His system of charting elements arranged them in rows based on increasing atomic weight (from left-to-right). The rows doubled back to form columns based on those elements within a column having identical ratios of bonding with other well-known elements. The resulting table not only explained much existing chemistry, but predicted some of the weights and combining prop-erties of elements which were yet to be discovered. Indeed his table caused some previously accepted atomic weights to be reinvestiga-ted and subsequently revised. Each new revision or elemental dis-covery confirmed the essential correctness of Mendeleev's effort, overcoming the fact that while Russia certainly had extensive expe-rience with mineral chemistry, it was not a major scientific center. Today, the work of this Russian is embodied in virtually every sci-entific classroom in the world in the "Periodic Table" (so called because the combining or reacting properties of some elements regularly recur with each column down). This table has importance even today for the literature of inorganic chemistry since some specialty journals categorize elements they emphasize by their groupings in the table.

## The Modern Concerns of Academic Inorganic Chemistry: Three-Dimensional Structures, Complex Bonds, and Coordination Chemistry

Two major categories of topics — coordination chemistry and organometallic chemistry — dominate today's agenda of inorganic chemistry. Both involve a characteristic stretching of what constitutes an inorganic compound.

The first and essentially older of the two subspecialties is coordination chemistry. To a substantial degree, coordination chemistry is the study of certain complex acids and bases, and their crystalline salts. Historically, it had been observed that in the most simple acids or bases there were two elements dissolved in a solution. One was regarded as a something of a negative ion (although the electronic connotations given these terms would come later) and another as a positive ion. Their behavior was predictable, and at least partly explainable by Dalton's conceptual scheme. Both in terms of illustrations in the journal literature of that time and in terms of the scientists' mental images of many scientists, most compounds were regarded as essentially two-dimensional, or "flat." But as research progressed in the late 1800s it became apparent that many acid-base compounds turned out to be highly complex, and projected rather prominently in three-dimensional space. They were really not "flat" at all. The qualifying factor that made this subject "inorganic" was that despite the fact that some of these compounds did have carbon in an indirect attachment to the center, a metal atom was always at the center of the every complex.

Further, the complex did not easily break down into individual elements when in solution. Rather, these surprisingly large and stable complexes behaved in many respects as if they were a single type of atom in solution. This was initially highly confusing to Dalton-inspired chemists. When dealing with solutions, they expected a natural breakdown to the level of individual elements and always sought explanations in the straightforward terms of the early atomic model.

Still another dilemma for strict Daltonists might be that not all of the complexes containing iron when dissolved in a solution, combined in exactly the same way that uncomplexed iron alone did when it was in solution. Most heretically, certain complexes might have iron binding to another atom in ratios of two sometimes, but with three at other times when two might be expected. This posed a theoretical problem of great interest for those who viewed the Dalton atom as essentially unchanging and of constant character with respect to combining or bonding to other compounds in fixed ratios. What was needed was a more sophisticated understanding of bonding.

Indeed, it has been the discovery of the multidimensional structures and the cataloging of the numerically varying bonding sites found in these coordination compounds that has driven inorganic chemistry for much of this century. This ongoing exploration has given inorganic chemistry another of its titles and expanding subject domains: "structural chemistry."

This effort has been highly international, and by the 1930s had effectively replaced (among chemists at any rate) the older competition to discover new elements. Alfred Werner, a scientist at the University of Zurich, won a Nobel Prize in Chemistry for work in coordination compounds in 1913, but American academics have also played important roles. G. N. Lewis of Harvard, MIT, and Berkeley suggested an early workable electronic theory of acids and bases that, while not quite correct in its finest detail, is still useful in teaching and in predicting many practical reactions. This was a tremendous advance when one considers that the electron, the atomic particle most important for chemical bonding, was not really experimentally measured until about 1918 by Millikan at the University of Chicago (Nobel Prize in Physics, 1923). Refinement in terms of details at the atomic level was due to Linus Pauling of Berkeley, Columbia, and Caltech, whose electronegativity series incorporated the latest European thinking plus his own insights. Pauling's modern ("quantum mechanical") theory of the chemical bond won him the Nobel Prize in Chemistry in 1954. (He also won the Peace Prize in 1962.) As if to confuse the reader further, Mulliken (with a "u" and an "e" — but, like Millikan, from the University of Chicago and also with a Nobel Prize [1966 in Chemistry]) is also famous for work on chemical bonding, but its impact was felt largely in what is now called theoretical chemistry, a field within physical chemistry and chemical physics. Arguably the most famous inorganic bond theorist of today is Roald (not Ronald!) Hoffmann of Cornell. He is also a published poet. Hoffmann won the Nobel Prize for Chemistry in 1981. He has not, as yet, received a second telegram from Stockholm for his verses.

### Aiding Its Seeming Rival:
### Inorganic Chemistry Empowers Organic Chemistry
### to Create Hitherto Difficult Organic Molecules
### in Abundance Through Organometallics

While there are hundreds of coordination compounds under study, there are thousands of organometallic compounds being pursued today. Organometallic chemistry is the largest component specialty within inorganic chemistry, and once again, represents a stretching of what is a proper domain for the field. Each organometallic compound contains, by definition, at least one carbon atom bonded directly to a metal atom. Like coordination compounds, organometallic compounds are often highly geometric — like a school-yard jungle gym or a geodesic dome — and have numerous interesting bonding arrangements. Two facts have made organometallic chemistry even more prolific than coordination chemistry. One has been its even closer relationships with the field of organic synthesis, the laboratory creation of new carbon compounds. The second is the fact that some organometallic compounds are even more useful than coordination compounds in medicine and chemical manufacturing through superior catalytic performance. Organometallics have maintained this edge despite the fact that many organometallic compounds are less stable than coordination compounds and sometimes require rather exotic methods of laboratory production and special attention to safety. (The famed Bunsen lost an eye to an explosion while working on them.)

Chemists were isolating and then using some organometallic compounds even before they understood just what they looked like molecularly. Prussian blue, an iron-carbon-nitrogen compound, was the dye of the uniforms of Frederick the Great's army. A number of toxic or medicinal compounds were known in the 1800s: Zeise's salt (work in Denmark, involving platinum), and Cacodyl, (an early antisyphilis agent from Germany, involving arsenic). But it was when organometallic compounds became useful for facilitating the step-by-step building of new compounds that the field truly gained impetus. At times progress would be made on substances, called intermediates, that served as useful building blocks. At other times, the new compound would serve as an accelerator, or catalyst of a desired series of reactions.

The catalytic successes were not unexpected. A number of simple metals and coordination compounds had long been discovered by inorganic chemists as useful in catalysis and these successes spurred the search for catalytic properties among the new organometallic compounds. Frankland of England developed a series of organo-nickel and organo-zinc compounds that sped along otherwise slow and uncertain organic reactions, while Grignard of France won a Nobel Prize in 1912 for the extreme usefulness of his organo-magnesium intermediates. Karl Ziegler of West Germany would win the 1963 Nobel Prize for 1930's work that included organo-aluminum catalysts in the processes we now use to make many plastics. Herbert Brown of Purdue took the 1979 Nobel Prize for his 1950s and 1960s work with many boron compounds including organo-borons, once again stretching the notion of what compounds were properly organometallic (boron is not commonly considered a metal) even as he was building up a new class of highly useful chemical intermediates.

Academic organometallic interests were tremendously advanced in the late 1950s and early 1960s when a remarkably stable organo-iron compound, ferrocene, was synthesized. Its synthesis was not all that arduous and it proved exceptionally tough while undergoing a variety of chemical and physical probes. Indeed, the simple elegance of its making caused its discovery to be considered as something of a parlor trick, and Geoffrey Wilkinson, a Briton teaching in the U.S., and its maker was denied tenure at Harvard. History would show that ferrocene studies greatly facilitated the basic understanding of organometallic compounds, and an appropriately crimson-faced Harvard was to soon to regret its decision when he won the Nobel Prize in 1973. Wilkinson, who had by then returned to his native England, also developed practical catalytic compounds, in his case using the less-common metal rhodium. Wilkinson is also famous as the coauthor of a series of encyclopedic texts in inorganic chemistry along with F. Albert Cotton, of Texas A&M University. Cotton is himself probably the most published organometallic chemist of all time, with over fifteen hundred papers to his credit.

### Bioinorganic Chemistry

Biomedical interest in coordination and organometallic compounds has been one of the less appreciated specialties, despite the fact that a large number of life substances now qualify as inorganic in a way that would surprise scientists of the 1700s and 1800s. Iron turns out to be central to hemoglobin, the key blood component, just as magnesium is important to chlorophyll, the key compound of green plants. Vitamin B-12 centers around a cobalt atom and serves a catalytic function as do trace amounts of many minerals such as zinc. Sulfur has a major role in framing long strands of nucleic acids and linking loops of the insulin molecule. Calcium and phosphorous are the major components of our teeth and skeletons. Inorganic chemistry has long provided us with poisons; what is less appreciated is that modern inorganic chemistry also has provided us with ways of snatching them out. "Chelate" compounds (from the Greek word for claw) are products of coordination and organometallic studies, and have had some success in removing the lead in the bloodstream of children who have eaten paint flakes.

### Crystallography for Chemists

In an historical context, modern biochemistry also owes a debt to inorganic chemistry owing to the inorganic chemist's preoccupation with structural matters, particularly the structures of crystals. Inorganic chemists long ago learned from physicists and mineralogists the power of X-rays to indicate structure, and have passed this on to chemists working with proteins and nucleic acids. Many complicated biochemicals, much like coordination compounds, can be crystallized for a much more rapid analysis using X-rays than would be possible using "wet" chemical tests. The various formulas for converting patterns on exposed X-ray films into coordinates for three dimensional views were developed by inorganic crystallographers. Indeed, two Americans, Hauptman and Karle, who were doing basic research in crystals of use to military electronics in the fifties and sixties, were to win the 1985 Nobel Prize for significantly simplifying the determination of crystals structures. Symptomatic of the flow of this information to the biological sciences, Hauptman has gone on to medical research administration.

## Inorganic Chemistry and Advanced Materials

Inorganic chemistry has also developed an interest born out of crystallography for solids that have quasi-crystalline, semiconducting, superconducting, or metal-like properties. Hybridized and composite materials of these types — including both natural and synthetic silico-aluminum complexes like zeolites — are of increasing importance in the chemical processing, ceramics, and electronics industries. Indeed it is highly likely that the car you drive has a platinum-composite catalytic converter. Areas such as these represent a tremendous growth field for inorganic chemistry.

## Nuclear Chemistry and Radioactive Materials

Conversely, the chemical study of radioactive materials, particularly isotopes, has been a province of inorganic chemistry that has declined somewhat over the years. Four earlier Nobel Prizes (Curie, 1911; Hevesy, 1943; Hahn, 1944; and Seaborg, 1951) represented a golden age of the discovery of exotic new elements or particularly scintillating radioactive versions of more ordinary materials. These materials remain important as tracers or labels of materials in the body. The metal thallium, for example, is injected into the bloodstream to be picked up by machines monitoring congested blood flow, typically in clogged coronary arteries. In other cases, such as overactive thyroid glands, radioactive iodine can suppress their function. Industrially, radiochemistry is useful in analyzing the incorporation of a compound, desired or undesired, by labelling that compound with telltale radioactivity, and checking the finished product with the equivalent of a Geiger counter. Environmentally, it is likely that the servicing of older nuclear reactors will sustain research into the effect of radioactivity on metals and concrete building materials.

## SORTING OUT THE JOURNALS

### Journals of General Inorganic Chemistry

There are perhaps a dozen journals of general inorganic chemistry worldwide. Historically the most important have come from the Continent. The *Zeitschrift fuer Anorganische und Allgemeine Chemie* (Barth) is representative of the German school, even as the *Revue de Chimie Minerale* (Dunod, Gauthier, Villars) historically reported French work. While both of these journals now accept papers in English, they have tended to lose the spotlight to more obviously internationalized or Americanized journals. The *Zeitschrift* still, however, features much mineralogical chemistry, and the *Revue* has very enterprisingly transformed itself into the *European Journal of Inorganic and Solid State Chemistry*, with results yet to be determined.

Today, two of the leaders come from exclusively English-speaking societies. The American Chemical Society sponsors *Inorganic Chemistry* while the British Royal Society of Chemistry publishes the *Journal of the Chemical Society, Dalton Transactions*. As shown in Figure 1 both have impressive impact factors, and do well in cost comparisons. (See Figure 2.) It is arguable that their standing would be even more impressive save for competition from within their own society's publishing houses for good papers.

In the U.S., Americans have two other major society outlets for good inorganic chemistry. A significant amount is reported in the *Journal of the American Chemical Society*. Further, the ACS apparently started up *Organometallics*, a rival sectional journal covering the booming field of organometallics without the complete assent of supporters of *Inorganic Chemistry*. Perhaps wisely for its own survival, *Inorganic Chemistry* has refused to relinquish the popular field — and indeed is starting to publish twice as often as *Organometallics*, using speed of manuscript publication as a powerful attract-

FIGURE 1

# Contributors and Relative Impact
# Leading Inorganic Journals

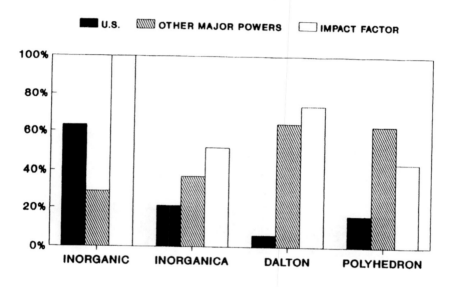

■ U.S.    ▨ OTHER MAJOR POWERS    ☐ IMPACT FACTOR

ant — so that both ACS journals now carry excellent organometallics papers. In Great Britain, there is only one society journal competitor: the Royal Society's *Journal of the Chemical Society, Chemical Communications*. Many short letters-type papers in inorganic chemistry appear there. Once again, quick turnover is an effective inducement.

Surprisingly given these four outlets, not all of the U.S. and U.K. output is accounted for. Outlets needed largely by inorganic chemists in Continental Europe are also yet to be considered. This mixed constituency is serviced primarily by two capably run competitors outside the chemical societies: *Inorganica Chimica Acta* from Elsevier and *Polyhedron* from Pergamon. *Inorganica* publishes about twice as many papers as *Polyhedron*, but at more than *Polyhedron*'s its cost. In terms of manuscript geography, *Inorganica* has had more American papers in both relative and absolute terms until recently. But overall *Polyhedron* has a higher, and more

FIGURE 2

# Comparison of Number of Papers, Costs, and U.S. Market Penetration

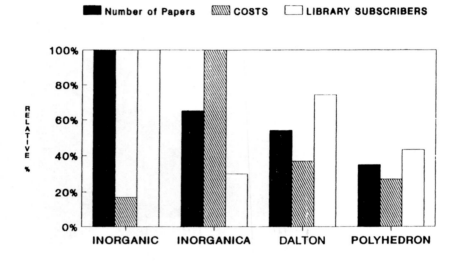

rapidly increasing share of papers from scientifically competitive countries, including the U.S. The difference in *Polyhedron*'s favor may be the involvement of Sir Geoffrey Wilkinson, arguably the world's most famous inorganic chemist, as editor in chief. *Inorganica* remains, however, a better choice for schools with a strong bioinorganic interest.

General inorganic chemistry has at least two traditional review journals: *Progress in Inorganic Chemistry* (Wiley) and *Reviews in Inorganic Chemistry* (Freund).

### Journals of Coordination Chemistry

While journals of general inorganic chemistry still publish the bulk of coordination chemistry papers, two journals specifically focus on this area. Gordon and Breach has published the *Journal of Coordination Chemistry* since 1971, with a stress on original research papers. Somewhat lengthier, overview papers are featured in Elsevier's *Coordination Chemistry Reviews*.

## Journals of Organometallic Chemistry

One of the financially most crucial decisions in inorganic collections is the depth of library commitment to organometallic journals. Two journals dominate the field. (See Figures 3 and 4.) One is undoubtedly very expensive, and its pricing has, in effect, spawned its own competitor. The American Chemical Society entry, *Organometallics*, was started in 1982 as a deliberate response to the cost (over $3,000 annually) of its for-profit competitor, the *Journal of Organometallic Chemistry* from Elsevier. This attempt included a raid on *JOMC's* considerable editorial talent, and *Organometallics* has shown much success with many U.S. subscriptions and a higher impact factor (see Figure 2). However, the *Journal of Organometallic Chemistry* has not folded its tent. Despite the fact that

FIGURE 3

# Contributors and Relative Impact Leading Organometallics Journals

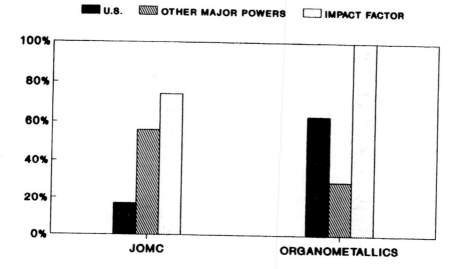

FIGURE 4

## Comparison of Number of Papers, Costs, and U.S. Market Penetration

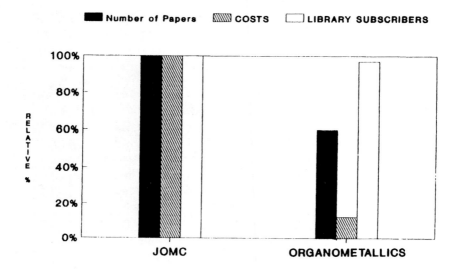

*Polyhedron* also started in 1982 and could be expected to siphon off even more good papers, *JOMC* has maintained a good reputation. It has a special strength in exhaustive review papers published at regular intervals covering most of the major groups of inorganic elements in the periodic table. *JOMC* remains essential for all graduate programs involving inorganic chemistry.

Another valuable source of reviews is the hardbound series from Academic, *Advances in Organometallic Chemistry*. These have a freer range of topics and make for a nice complement to those reviews in *JOMC*.

### Journals of the Structural Chemistry of Crystals and Other Solids

The most important structural studies of inorganic and organometallic interest appear in two categories of journals of solids. The distinctions between these two categories of subject journals—journals of "crystallography" vs. journals of "solid-state chemistry" — are not absolute, but a few guidelines follow:

- On the whole, purely crystalline solids and structural studies of a single compound tend to be favored in journals of crystallography. There have been many of these "crystallography" journals for many years.
- By contrast, solids with a mixed or composite structural character, or studies comparing several differing structures for differing chemical or electronic effects, tend to be reported in a smaller and younger assortment of "structural chemistry" journals.
- While the older generation of crystallography journals formerly had enormous involvement from the physics community in terms of the application of X-rays to elucidate complicated structures (from the 1915 Nobel Prize work of the father-son team of W. H. and W. L. Bragg to the 1985 Prize for Hauptman and Karle) most of their current audience are chemists.
- By contrast, the younger generation of structural chemistry journals still involves a great many physicists, largely because the solids studied have potential in electronics or interest for solid-state physics theory.

As seen in Figures 5 and 6, three of the leading journals specifically refer to crystalline materials: *Acta Crystallographica B, Acta Crystallographica C*, and the *Zeitschrift fuer Kristallographie*. The *Zeitschrift* is the most general of all in terms of stated scope, although it has had a traditional focus on reporting structures rather than on the chemical effects of variant structures. *Acta B,* on the other hand, is subtitled *Structural Science*, and in many respects is most comparable to the *Journal of Solid-State Chemistry* in its wider range of studies and focus on the chemical effects of more than one structural variant of a crystal. *Acta C* is subtitled *Crystal Structure Communications* and is probably the world's largest repository of individual structure determinations.

(There is, of course, an *Acta A*, subtitled *Foundations of Crystallography*, but it has a pronounced mathematical and basic physics flavor. It deals substantially with derivations and formulas for de-

FIGURE 5

## Contributors and Relative Impact
## Leading Crystal & Solids Chem. Journals

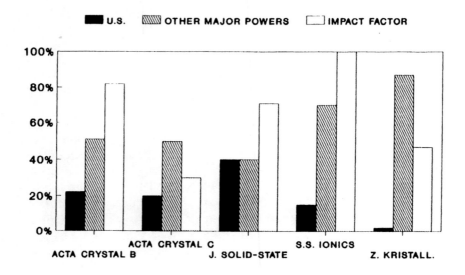

termining structures. Chemists are interested in *Acta A* more for the occasional improvements in determination methods reported for their particular category of crystals, rather than as a daily companion to their studies.)

While the *Zeitschrift* might appear to be a ready first choice for smaller collections because of its broader scope and the high number of papers that it publishes, certain factors militate against it. Its price is higher than that of any one of the *Acta* sections and it remains intensely Central European in its author constituency, not surprising in light of Oldenburg, its entirely reputable but largely localized German publisher. While *Acta's* publisher is also European (Munksgaard of Denmark) all sections of *Acta* have a significantly higher proportion of American papers and even a wider variety of papers from European countries. *Acta C* is probably the best single choice when only one journal is possible. It simply reports

FIGURE 6

## Comparison of Number of Papers, Costs, and U.S. Market Penetration

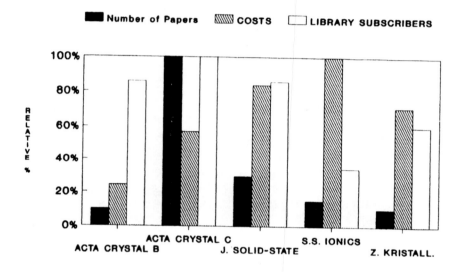

more new structures of interest to chemists, and that straightforward interest in structure is a hallmark of the field.

Another journal, not depicted in our graphs, that might be of interest is the *Journal of Crystal Growth* (Elsevier, 1967). While this title formerly reported a good deal of crystal growth from wet solutions—the manner of traditional inorganic chemistry—it has more recently focused on growth from molten materials or materials processed in other ways specific to the semiconductor industry. All manner of crystal growth and characterization is covered, however, by the review journal *Progress in Crystal Growth and Characterization* (Pergamon).

The *Journal of Solid-State Chemistry* and *Solid-State Ionics* are comparable in that both take not only crystalline but noncrystalline and composite materials as well. The *Journal*, not surprisingly, is more chemical than physical in interest than *Ionics*. But *Ionics* tends to treat many materials that could be regarded as similar to acid-base salts, an area of long-standing interest to inorganic chemists. Nonetheless, if one is forced to take only one of these, the *Journal* is a better choice. It has more papers at a lower cost, and has a heavier U.S. author involvement. The balance is tipped in the opposite direction when a very strong physics or electronics community is also to be served. In the latter case, the *Journal of Physics and Chemistry of Solids* (Pergamon) might also be considered. *Progress in Solid-State Chemistry* from Pergamon is the leading review journal in this area.

### Journals of Chemical Catalysis

The study of ways of speeding up the bulk transformation of chemicals is one of the principal themes in virtually all chemical engineering journals. Support for inorganic chemists working with catalytically active elements or metal-composite systems is tremendous from the chemical processing industries. While a comprehensive collection serving catalytically concerned inorganic chemists should have some chemical engineering titles, most of the more academically concerned scientists emphasize three journals as outlets. (See Figures 7 and 8.) While the *Journal of Molecular Catalysis* and *Applied Catalysis* present an even finer distinction of differences between the pure and industrial segments of this community, the *Journal of Catalysis* serves both segments in the U.S. somewhat better, and at lower cost. This is probably due to its American publisher, Academic (vs. Elsevier) and head start (1962 vs. 1975 and 1981). Nonetheless, serious collections should strive for all three of these fine titles, and perhaps *Catalysis Letters,* another Elsevier title that has substantial foreign involvement.

Another pertinent title, but of a more specialized nature, is *Zeolites* (Butterworth). Zeolites, mentioned earlier, are a special category of solids with a large cage or sieve-like structure that allows smaller compounds to flow fairly freely through them. Some zeolites can be structured so that catalytic metals are embedded in the sieve. Reacting chemicals that flow through such enhanced sieves experience faster or more complete transformations. These make zeolites especially useful to catalytic scientists, who have written up

FIGURE 7

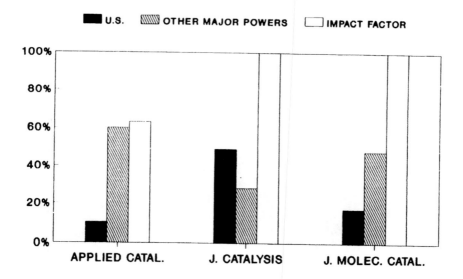

# Contributors and Relative Impact
# Leading Chemical Catalysis Journals

hundreds of accounts of both naturally occurring and synthesized zeolites.

Review papers in the overall field of catalysis are provided by *Advances in Catalysis* (Academic) and *Catalysis Today* (Elsevier). The former title is more formal and in-depth, the latter more current and informal.

## Journals of Nuclear Chemistry and Radioactive Materials

This area of diminishing interest among U.S. chemists remains surprisingly well populated with titles from abroad. Oldenburg, already noted for *Zeitschrift fuer Kristallographie*, is represented by *Radiochimica Acta*. Once again, this title has the broadest scope in its class and while again heavily European, attracts somewhat more American papers than its competition.

FIGURE 8

# Comparison of Number of Papers, Costs, and U.S. Market Penetration

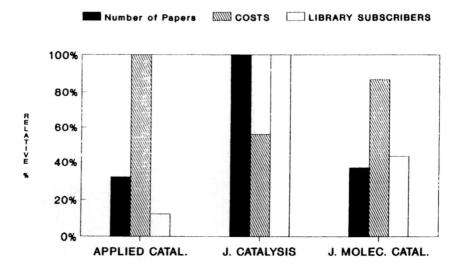

Slightly more specialized is an Hungarian-based Elsevier entry, *The Journal of Radioanalytical and Nuclear Chemistry*. This title is, however, not much frequented by American authors, and unless analytical chemistry is stressed in the collection, it remains a secondary choice.

More American authors appear in *Applied Radiation and Isotopes* and the *Journal of Labelled Compounds and Radiopharmaceuticals* (both from Wiley), but the inorganic chemistry content in these journals is fairly small. They do remain a good choice for medically oriented collections.

The *Journal of Nuclear Materials* (Elsevier) serves the chemical aspects of nuclear reactor operations well, although once again, its constituency is primarily among engineers, not mainstream inorganic chemists.

The interactions of radiation with solids, particularly with crystals or composites of importance to electronics, is handled well by *Radiation Effects and Defects in Solids*. The primary audience of this journal in the U.S. scientific community, however, is among solid-state and low energy nuclear physicists.

## Journals of Bioinorganic Chemistry

This growing field, in ironic contrast to radiation chemistry, is served by surprisingly few journals specifically devoted to it. The *Journal of Inorganic Biochemistry* (Elsevier) is the leader. Its greatest competition comes from papers in two more general inorganic journals: *Inorganic Chimica Acta,* which has long welcomed bioinorganic papers, and *Inorganic Chemistry*, a late-comer to bioinorganics. *Inorganic Chemistry* has deliberately sought to expand the numbers of papers in this area since the arrival of its in-house competitor *Organometalics* threatened to undercut its support in organometallics. Certain sections of the titanic Elsevier title, *Biochimica et Biophysica Acta*, particularly the *Protein Structure and Molecular Enzymology*, report a good deal of bioinorganic chemistry in that many enzymes feature a central metal atom in their structure.

The role of trace metals (zinc, iron, calcium, etc.) in health and toxicology is often found in other journals of specialized biochemistry or in journals of nutrition and pharmacology. Examples include *Biological Trace Element Research* from Humana, *Magnesium* from Karger, and *Cell Calcium* from Churchill Livingstone. Dekker publishes a review series, *Metal Ions in Biological Systems*, that is particularly pertinent.

## Journals Covering Specific Categories of Inorganic Elements

Journals of interest to inorganic chemists are arrayed not only by thematic or applications orientations, but according to their chemical families as seen in the periodic table.

The oldest of these titles is the *Journal of Less Common Metals* (Elsevier). These metals are found in two rows of the periodic table called the actinide and lanthanide series. The title *"Less Common"* is something of a misnomer for two reasons. First, geochemists now recognize that some of these "rare earths" are much more common than was previously recognized. The historical problem was really one of isolating them in their pure forms. Second, the seemingly esoteric tone of the title masks the fact that these metals have already had three eras of strong scientific and engineering interest. Initially, rare earths were used to specially reinforce steel alloys. Then, they were used to coat the interiors of television picture tubes because they luminesced in vivid colors. Now, it turns out, mixtures incorporating yttrium, ytterbium, and lanthanum, members of this family of "less common" metals, have been shown to have high temperature superconductivity, making the inclusion of this title in solid-state physics and electronics collections almost mandatory in addition to any reasons of chemical interest.

The transition metals occupy the center of the periodic table, essentially bridging the gap between the light alkali metals (e.g., lithium, calcium, and magnesium) and the halogens (e.g., fluorine, chlorine, bromine). The transition metals include some of the most economically important metals like iron, copper, zinc, gold, and silver. Most of these metals also form chemically interesting coordination and organometallic compounds. *Transition Metal Chemistry*, from VCH of West Germany, is then an essential title in any large inorganic collection.

While metalloid topics are dominant within inorganic chemistry, at least two other titles relating to nonmetals must be mentioned. The *Journal of Fluorine Chemistry* (Elsevier) deals with all aspects of this halogen in its gaseous and liquid forms, in its elemental and compounded state, and in both its academic and applied aspects. In a similar manner *Phosphorous, Sulfur, and the Related Elements*, a Gordon and Breach title, deals with elements and compounds not typically regarded as metals. Since arsenic and selenium are also members of columns V and VI of the periodic table, work on them is also included in this journal.

# Chapter 3

# Organic Chemistry and Its Journals

## BACKGROUND

Organic chemistry is the study of carbon compounds. It has been one of the dominant specialties within chemistry for almost 150 years. Organic chemistry has been especially favored in that carbon is the main element in an overwhelming number of essential substances. Carbon is the principal constituent of the living tissues of most plants and animals. It is also the most important ingredient in most fuels, medicines, dyes, and rubbers. The carbon atom is tremendously amenable to bonding with itself and other atoms in synthetic compounds of an almost endless variety of shapes.

### Could Compounds from Nature Be Duplicated?

Chemists of the early 1800s initiated a two-step dance of organic chemistry that has continued ever since. The first step was extracting new chemicals from plants, animals, or carbon-containing minerals. The second step was then confirming the composition and structure of the extract by creating a duplicate in the laboratory using known ingredients. Early chemists had long been better at the first step—the extraction of compounds from natural materials. Indeed, many early chemists believed that the second step—duplicating these "organic" compounds in the laboratory—was an impossibility. This belief was shattered when Friedrich Wohler, a German chemist, was to report with glee: "I can no longer, as it were, hold back my chemical urine . . . I can make urea without a kidney, whether of man or of a dog!"

Wohler was extremely fortunate in that one of his principal champions was Justus Liebig, an immensely forceful academic compatriot. Liebig was trained initially in pharmacy, which was then largely a practical trade of plant and animal extracts. But Liebig sought greater intellectual challenges than a druggist's career promised. With the aid of grants from the local nobility, given to him despite his extreme contentiousness in both political and scientific matters, he received higher degrees in more formal chemistry in both Germany and France. This combination of practical experience (he insisted that all his students personally work weekly in a real laboratory—a novelty for many of the gentility that attended universities and were accustomed to having their experiments done by servants) and theoretical training helped make him, without question, the greatest natural products chemist of the nineteenth century. Despite having invested his own life's work in extractive chemistry, especially in topics of interest to agriculture, Liebig was not in any way defensive about the notion or significance of purely synthetic organic compounds. He had done a good deal of inorganic synthesis early in his academic career, and expected that organic synthesis would one day come. Liebig wholeheartedly used his extensive program of training new chemists and his editorship of the influential *Annals* to strengthen belief in the reality and potential of synthetic organic chemistry.

## The German School of Organic Chemistry

By the early 1840s a new incentive for organic chemistry emerged. It started with two of Liebig's students, Ernest Sell and August Hofmann. After graduation, Ernest Sell had gone on to run a coal processing factory. Sell sent a sample of a coal tar extract to Hofmann, who was continuing his studies. Work with this extract led to the discovery of the anilines, a family of organic compounds of remarkable versatility. Many of the anilines could be manipulated synthetically in the lab to produce a broad range of colors of use to the dye industry (yellow and green were among those initially discovered by the Germans).

Adolph von Baeyer, another of Liebig's students, furthered this theme by his work on synthetic indigo dyes. Indeed von Baeyer is essentially the source of the tint of today's blue jeans. He also discovered a large number of compounds of medicinal importance, including the barbiturates. Von Baeyer won the Nobel Prize for his work in 1905.

Liebig's influence went beyond the German-speaking countries, and England indirectly owes its school of organic chemistry to him as well. The same Hofmann mentioned earlier gave a series of lectures in England that attracted William Perkin, the practical son of a rather hardnosed local building contractor. Young Perkin was assigned the laboratory duplication of quinine, the antimalaria drug extracted from Peruvian cinchona bark. He thought he had made a serious mistake when he came up with a purple substance that did not resemble natural quinine. He nonetheless hit upon the practical idea that it might be a good dye and sent off samples to Britain's textile manufacturers, then the world's leaders in textiles. His business instincts were right. (In fact, although the German group discovered other anilines first, they were beaten to their first practical application by their pupil Perkin!) Perkin was able to retire at age thirty-six to pursue a long career as an independent researcher, funded by the sale of his dye business. Together with his son, William, Jr., also a renowned chemist, and his son's best friend, Frederick Kipping (who like Perkin's son also went to Germany to study under von Baeyer before returning to England), they effectively became the nucleus of Britain's efforts in organic chemistry.

Liebig's students established research centers in organic chemistry in the U.S., Canada, and virtually all of the European countries. France and Belgium, however, had a series of leaders who developed independently from a tradition that was strong before Liebig had seized the attention of scientific Europe.

## The French School of Organic Chemistry

Despite the fact that Liebig had trained in France, he repeatedly stoked the flames of Franco-German rivalry in organic chemistry. Since the time of Lavoisier, French chemistry had been regarded as the world's most advanced, with a galaxy of academic stars. In fact, until the rise of Liebig, Dumas of Paris was regarded as the world's foremost authority in organic chemistry. (A prominent Swedish chemist, Jakob Berzelius, the widely read editor of the first annual review journal, actually announced rankings concerning the relative merits of given chemists active in his day, and his ratings were very much respected and feared.) Ironically, the career of Dumas was very much like Liebig's. Like Liebig, Dumas came to organic chemistry through pharmacy. He had dozens of famous former pupils, much like Liebig. Dumas, too, was supported well by the government, (he was, by contrast with Liebig, however, quite a successful politician). Like Liebig, Dumas also edited an influential chemistry journal. (Indeed the journal that Dumas edited was also entitled *Annals* and thus began the headaches of many a librarian chasing down references to two journals in the same field with similar names!) There the similarity ends, for Dumas was an especially agreeable person, while Liebig was notoriously tactless.

Liebig came to feel that Dumas' disagreements with him on matters of chemistry were, in fact, rooted in French national arrogance. A basic theme of Dumas' work was that the laws, terms, system of atomic weights, and classification of substances used in chemistry ought to be as all-encompassing as possible. This was certainly a noble idea and had a long-range validity. It was entirely in line with the thinking of the leading English chemist of the previous generation, John Dalton. Liebig, however, disagreed violently in the details of Dumas' scheme at almost every possible juncture. Liebig had seen that there were fundamental differences in the bonding of inorganic versus organic compounds, and felt the attempts at generalization by Dumas were premature. The fact that Liebig was generally right scientifically in the individual details has obscured the fact that Dumas was otherwise quite successful as a practical organic chemist. Dumas was among the first to isolate the many different varieties of chemical alcohols and ethers and, after Wöhler's triumph, Dumas went on to make many of them from scratch. Ultimately, however, Dumas incurred the wrath of Berzelius as well, and retired from active research to pursue politics full-time.

Auguste Laurent was the most influential of Dumas' pupils. He made a small fortune working in the burgeoning French perfume industry, another fertile source of organic chemists. Much like Perkin's situation, this enabled him to devote a good portion of his middle years pursuing whatever chemistry interested him. His working life exhibited the classic pattern of early organic chemistry. He first isolated natural perfume products and then worked on synthetic versions. But Laurent was one of the first French chemists to realize that attempting to study hundreds of widely different organic compounds at once and to quickly come up with rules of reactions and classification schemes valid for all of them was suicidal. In this recognition of personal and technical limitations, Laurent differed from his mentor, Dumas. Laurent settled in narrowly and specifically on the naphthalene family of organic compounds (strong-smelling substances derived from coal and today used largely in dry cleaning). He systematically explored every reaction possible to him within that restricted group and made careful, qualified proposals as to what his experiences revealed about organic chemistry in general. While he did not have the whole picture about even the naphthalenes at the time, because certain concepts of structure and bond-sharing were not yet known, he developed out of his own experiences widely adopted concepts and terms. Words and phrases like "radical," "substitution reaction," and "isomer," that form the everyday language of working organic chemists today, either originated with or were most accurately furthered by Laurent's articles and books. He salvaged what was useful in Dumas' classification scheme. Indeed, Laurent was one of the few scientists of the age who was not afraid either of the elderly Berzelius nor of the rising young Liebig.

Jean Stas was a Belgian who came to work with Dumas after having received a medical degree and having experience with the chemistry of fruit tree saps. He enjoyed some fame in his time for atomic weights and chemical theories that he developed with Dumas in opposition to Liebig. However, his most lasting contribution to organic chemistry was the application of all of his varied medical, plant, and chemical experiences to the first serious study of nicotine, the principal toxic component of the tobacco plant. (Indeed, he testified at a famous murder trial in which he was able to extract nicotine from the corpse.) This work had serious long-range importance for both medicine and organic chemistry. His methods better enabled the detection and isolation of a large family of potent drugs known as the alkaloids, of which nicotine is a member. (Heroin and strychnine are other members.)

The generation of French chemists after Dumas was greatly indebted to Dumas intellectually but somewhat less consumed with the Franco-German rivalry in chemistry. Indeed, they extended an olive branch to the rival field of inorganic chemistry as well. Grignard, as noted in the inorganic chapter, developed a highly useful series of organo-magnesium synthetic building blocks, while Sabatier developed metal catalysts to speed along otherwise difficult organic reactions. Both men were to win the 1912 Nobel Prize in Chemistry, and their success represented the triumph of practicality in organic chemistry over allegiance to nationalist schools of chemical dogma.

## Organic Chemistry Grows Big Enough to Divide, This Time Not by Nations, but by Topical Specialty

Once it had been firmly established that organic synthesis was more than just a parlor trick possible for a few substances, and that the chemistry allowed by the nature of the carbon bond was truly special, organic chemists began to multiply in academic institutes and in industry everywhere. As the field grew, and the old Franco-German enmity died, organic chemists began to associate not so much with their chemical countrymen as with other organic specialists of any nationality, particularly with those who shared their subspecialty interest.

By the turn of the century, thousands of articles in organic chemistry appeared in journals of general chemistry. General purpose national chemistry journals began to have the problem of saving space for other nonorganic specialties. This pressure to limit the number of organic papers in already-established journals, and the natural urge for special recognition and control of their own journal outlets, led organic chemists to establish journals specifically devoted to organic chemistry or to its branches. The total expulsion of organic chemistry from general journals was, of course, unthinkable given the huge numbers of organic chemists. Even after the founding of specially tailored journals, organic chemistry tended to remain dominant in general chemistry journals.

## The Lords of the Rings:
## The Study of Cyclic Structures in Organic Chemistry

One of the first new specialties in organic chemistry was based on a structural novelty. Many carbon compounds, particularly those derived from coal, turned out to have closed rings of carbon, a type of structure never before seen in chemistry. A number of historical figures are important in the development of this academic specialty.

The first was August Kekulé, yet another of Liebig's German students. He was different from many chemists of the age in that he came to chemistry not from work in pharmacy, plant science, or mining, but from an early interest in art and architecture. Like Hofmann, he obtained employment in England. He began to wonder how it was that benzene, a stable coal extract, could have so few hydrogen atoms bonded to carbon atoms when it "needed" to have approximately double that number of hydrogens to use up all the attachment sites. He effectively asked himself: how could the carbons within benzene not be "attached" to something and still have benzene a stable compound? Compounds with "unattached" bonds were known to keep on combining with other substances—whether the chemist wanted them to do so or not—until all the bonds were attached. One day while riding on the upper level of a double decker bus (other versions have him gazing into a fireplace) Kekulé experienced something of a vision in which the carbon atoms danced in a ring. He hit upon the idea that the carbons could attach to each other in a ring, using up the extra attachment sites. He found that in some situations, carbon rings featured single, highly stable bonds between adjacent carbon atoms (which chemists had assumed was always the case), but that in other cases, rings could have somewhat less stable double or triple bonds, or even a mix. Indeed he was to demonstrate that many of the most aromatic materials, including Laurent's family of naphthalenes, were in fact systems of linked benzene rings with mixed single and double bonds between adjacent carbons. For a time it was thought that the practical notion of aroma was necessarily tied to the benzene type of bonding arrangement. But today, carbons with such an arrangement, whether odiferous or not, are said technically to be aromatic.

Kekulé went on, using his natural artistic-architectural bent to devise a number of other solutions to organic chemical quandaries based on the existence of rings with numbers of carbons up to eight. (Effectively, the shape of a "stop sign.") In another sign of the reconciliation of the French and German schools, Stas, the prominent student of Dumas, used his influence with the government to obtain for Kekulé a more prestigious position in the French-speaking universities in Belgium. There, Kekulé devised the graphical symbols which organic chemists use even now to denote ring structures, a work highly complementary to the development of the organic chemistry vocabulary of Laurent.

While Kekulé did not live long enough to win the Nobel Prize, one of his students did so in 1910. Otto Wallach demonstrated that these ring or "cyclic" compounds occurred not only in extracts from lifeless coal or in synthetic compounds but in many living plants, reinforcing the first step of the old "two-step" dance. The family of ring compounds he discovered is called the terpenes. They are based primarily on five-membered rings and do not typically have multiple or mixed bonds between adjacent carbons. Nonetheless, terpenes were shown to be primarily responsible for the odors of dill, caraway, Earl Grey tea, and citronella candles.

In 1939, Leopold Ružička, who worked in both Switzerland and the Netherlands, won a Nobel Prize for his demonstration that rings could have as many as seventeen member carbons. (Today certain common ether rings have up to eighteen carbons, and exotic rings of thirty have been made for purely scientific interest.) Some of these larger fifteen and seventeen member "macrocycles," as they came to be called, were isolated from the scent glands of small animals. Today, many of these fifteen and seventeen member rings are produced synthetically and are the essential ingredient in musk perfumes and after-shaves.

Over time it was realized that it was possible for rings to include one or more noncarbon atoms, usually oxygen, nitrogen, or sulfur. The inclusion of these different atoms within a ring causes it to be termed "heterocyclic." Today, in fact, heterocyclic compounds are dominant in cyclic studies.

## Biochemistry Begins to Diverge from Organic Chemistry, Leaving Behind the Bioorganic Subspecialists

Around this time those who worked in natural products extraction tended to split into two significant groups. Those who began to look at which isolates seemed necessary for the survival, growth, and development of living organisms began the field of biochemistry.

At first, biochemistry was indistinguishable from organic chemistry, excepting that it drew an increasing number of devotees from medical schools. Over time, certain categories of purified organic extracts came to be categorized structurally (as carbohydrates, lipids, nucleic acids, or proteins) or functionally (as catalytic "zymases" or, more commonly today, as enzymes). Members of the nascent biochemical fraternity seized upon these categories as their own "biochemicals" and it was not long before the Nobel Prizes came in. Emil Fischer, a student of von Baeyer, took one in 1902 for his work in sugars. (Emil's son Hermann became Berkeley's first professor of biochemistry, another case of Liebig's influence, one step removed.) Eduard Buchner, another of von Baeyer's students, took the 1907 prize for his work on enzymes.

A distinguishing feature of the biochemical school was that over time, the majority of biochemists showed less interest in duplicating these increasingly large compounds artificially through a total synthesis. Nor did they try to alter their components or structure to obtain different effects.

By contrast, a minority within the community came to be called "bioorganic" chemists. They attempted to perform many traditional organic procedures on the new categories of materials. They worked at reacting biochemicals with other chemicals. They tried molecular rearrangements and substituted components. At least in the early years, bioorganic chemists still studied structure by traditional "wet" chemistry methods. Straightforward biochemists, by contrast, began to use other "instrumental means" like chromatography and crystallography, to determine the size and structures of the larger compounds that concerned them.

Bioorganic chemists saw the large biochemical molecules as building-block assemblies of more fundamental organic units that could be disassembled and rearranged. Biochemists, in sharp contrast, sought techniques that kept the large molecule together and functional during the many steps it took to isolate it from interfering compounds during extraction and purification procedures.

## *Natural Products Chemistry Continues to Underwrite the Development of Medicinal and Academic Synthetic Chemistry*

In contrast to the medical school alumni that dominated the early biochemical group, the synthesis-oriented group attracted most of its adherents from pharmacy (a profession whose training was rather detached from medical schools) and from those trained exclusively in academic chemistry. The synthesis group deliberately worked to isolate compounds that were highly novel or intriguingly complicated. They viewed their total synthesis and modifications as challenges.

Pharmaceutical firms and university departments of chemistry — the latter often heavily subsidized by pharmaceutical firms — served as the major employers of this group of synthetic and natural products chemists. This group felt that they could not wait until the biochemists achieved a complete understanding of the life's work of a compound — a notion with which they were sympathetic but which seemed too distant scientifically to promise an immediate payoff. Rather, professors of pharmacy and organic synthesis worked on substances known through folklore to be effective for some ailment. Overwhelmingly these folklore remedies came from the medicinal plants in the botanical pharmacies of the Old World or from specimens brought in by explorers or anthropologists from the New World.

The payoffs with synthesis and careful chemical manipulation of these raw materials, which had at least some known therapeutic benefit, seemed more obvious and would come more quickly. In some cases, the payoff of performing the synthesis in the lab would be increased purity of the product, with fewer side-effects from unwanted contaminants in a crude extract. At other times, the natural source of the raw material would be rare or expensive, or the monopoly of some foreign country or competitor. Making the product synthetically often diminished those constraints. It also proved possible to improve on nature. Changing or rearranging important pieces of a naturally occurring molecule often increased its potency, or made it capable of being dissolved into a syrup or stable enough to be incorporated into a pill or capsule.

Of course, mistakes were made along the way. Misadventures on the way to a substance like today's methadone (which helps counteract narcotics addiction) included the synthesis of heroin and morphine to replace opium, both of which turned out to be more, not less addictive. An even more potent drug than these, Bentley's compound, turned out to be useful only for anesthetizing giant wild animals, like elephants on wildlife preserves.

Specialists in tinkering with compounds to obtain more desirable clinical effects are typically referred to today as medicinal chemists. Even today, medicinal chemistry, bioorganic chemistry, and natural products chemistry are closer to organic synthesis than they are to biochemistry. Biochemistry is a surprisingly late arrival to the pharmacological scene, and biochemical pharmacology tends to be featured largely in journals specifically devoted to itself, and rarely in journals of medicinal or organic chemistry.

Successes in natural products and pharmaceutically related organic synthesis were financially tremendous. Additionally, the insights derived from practical synthetic work brought purely academic organic synthesis up to the level of an art form and greatly increased an appreciation of cyclic chemistry. Nobel Prizes were awarded for work with steroids (Wieland, 1927 and Windaus, 1928), vitamin C (Haworth, 1937), vitamins A, $B_2$, E, and K (Karrer, 1937), hormones (Butenandt, 1939), alkaloids related to morphine and strychnine (Robinson, 1947), vitamin E and the active ingredient in marijuana (Todd, 1957), quinine, cholesterol, and cortisone (Woodward, 1965), and much more. Along the way it was discovered that many useful drugs (penicillin, streptomycin, Valium, Librium, and most barbiturates), virtually all sugars, and some important dietary amino acids (tryptophan), and insecticides (pyrethrin) were examples of heterocyclic chemistry.

The U.S. has become a full partner in the work of synthesis. **Roger Adams of the University of Illinois,** and Louis Fieser and Robert Woodward, both of Harvard, played the role of American-born founding fathers for serious organic work in this country (although all of them also had training in Germany as well.) In this generation, there are many stars, including Stork of Columbia, Corey of Harvard, Sharpless of MIT, Cram of UCLA, and Danishefsky of Yale. Leo Paquette of Ohio State has taken cyclic chemistry into another dimension: he has synthesized dodecahedrane, essentially a giant sphere, as well as several hundred other compounds, many of them heterocyclic. Yoshito Kishi of Harvard, Carl Djerassi of Stanford, and Alex Nickon of Johns Hopkins continue the classic interplay of natural products isolation and laboratory synthesis.

### Structures Are Not Created Equally Stable, Reactions Are Not Always Smooth: Physical Organic Chemistry

Organic chemists as far back as the Dutchman van t'Hoff and his French partner Le Bel (1874) had realized that conceptualizing carbon bonds as being flat often gave poor explanations of what was really happening. While it is quite true that most of Kekulé's and Laurent's aromatic ring structures were actually flat, a great many ordinary, nonaromatic bonding patterns could be best imagined as mounted on the points on the ends of a pyramid.

The appropriate pyramid would not, however, be quite like the Egyptian pyramids. Those pyramids are essentially triangles mounted on a square. These pyramids would be triangles mounted on a triangle, the resulting structure being more properly called a tetrahedron. Indeed, the purest form of carbon, a diamond, was shown to be essentially a crystal of packed tetrahedrons.

There had long been hints that organic molecules with basically the same member atoms but slightly different arrangements in three dimensional space behaved differently. A good example of such a structural difference comes from comparing one's left and right hands. Both hands have the same basic structure and yet, one cannot lay one hand on the back of the other and have the hands match. The thumbs are pointing in opposite directions. This is an example of a difference of spatial orientation leading to "stereoisomers." While literally hundreds of organic chemists have made contributions to this area, Ernest Eliel (then of Notre Dame) and Norman Allinger (then of Wayne State University) began one of the most systematic expositions of the problems in an influential hardbound review journal, *Topics in Stereochemistry*, and are themselves leading lights.

Another difference was noticed in some of the nonaromatic rings. If the rings were conceptualized edge-on, it would be seen that the accepted tetrahedral bonds would form jagged or sawtooth patterns. Differences in versions of that ring pattern came to be described as "conformational" and it was surmised, but not proven, that different ring conformations would lead to different properties or reactivity. To a substantial degree these problems were not fully solved, even for the common six-membered ring, until the 1969 Nobel Prize-winning work of Derek Barton of the U.K. and Odd Hassel of Norway. Conformational effects in other ring systems remain an ongoing research interest.

Ironically, early insights into conformal differences came from someone trained in chemistry, but much more famous in biology — Louis Pasteur. Pasteur put two solutions of sugar that had the same composition of atoms but, puzzlingly, were known to taste quite different, into shallow, clear, glass dishes. He then exposed them to sunlight. One type of sugar solution bent the light that went through in one angle, the other in a different, almost opposite direction. Pasteur knew that these sugars had cyclic structures. Given that they had the same atoms, he concluded that there had to be some difference in the design of the rings. He found later that he could make distinctions between other kinds of compositionally similar, but architecturally different, compounds using the same sort of simple light deflecting experiments. But Pasteur essentially got "sidetracked" into what became his life's work — microbiology. Full exploitation of Pasteur's chemical work was to wait one hundred years. Carl Djerassi, one of America's foremost natural products and synthetic chemists of the 1980s, became initially famous for his 1960s "rediscovery" of the structurally diagnostic use of light bending behavior in largely cyclic compounds — optical rotary dispersion — in a way highly reminiscent of Pasteur's work.

The issue of the "favoritism" that nature seems to confer on one spatial version of a compound over other spatial versions of the same compound that would seem just as suitable has sustained many chemical theorists. Over the early years of organic chemistry, scientists couldn't help but notice that spatially-oriented versions of a compound were significantly more numerous than their opposites. It was only when a detailed consideration of the location and relative sharpness of the angles of the bonds between the atoms was undertaken that some spatial arrangements were shown to be "strained" or "unstable." Unstable or strained versions, particularly of rings, were more likely to break down into simpler compounds over time. At other times, the unstable versions combined with other substances in a flask so quickly that they lost their unique identity. Sometimes when organic chemists sought to deliberately isolate or produce the less stable form, it would show a mind of its own, and start converting to the stable form, reducing its population in the beaker. Understanding relative stability has helped explain why some final products were favored over others when an isolation or reaction was complete. In order to get interesting reactions, stability is often something that has to be overcome. Sometimes reactions with stable compounds will not proceed smoothly without extra energy or use of a catalyst. Sometimes the reaction can be forced along to a desired end product, but requires proceeding along a somewhat different route of intermediate steps than might be expected.

Taken together, these emphases on structure-stability, structure-reactivity, spectra-structure, rates of reaction (kinetics), and routes of reaction (mechanistic studies), are generally considered physical organic chemistry. The details of all the topics that constitute physical chemistry will be dealt with later in their own chapter. Let it suffice for the reader that physical chemistry and organic chemistry have created a hybrid field—physical organic chemistry—which is closer to organic chemistry, and has generated a number of specialty journals in organic chemistry.

### The Ultimate Organic Chain Letter: The Rise of Polymer Chemistry

Polymer chemistry is essentially the study of extremely long molecules composed largely of repetitions of the same simple carbon compound building blocks linked end-to-end. Four phenomena tended to underwrite the development of polymer chemistry.

First, many natural products chemists failed to resolve certain natural gums, waxes, and resins into neat molecules with a definable beginning or end. This prompted a few of the most stubborn organic chemists to devote their lives to "attempting the impossible" by continuing to study them.

Second, synthetic chemists often came up with thick, viscous liquids, elastic films, or spongy solids, all of which they tended to view as failures or contaminations on their way to the synthesis of some elegantly clear or crystalline compound. (Clear crystalline compounds were often considered the hallmark of the successful creation of a pure new substance.) The urge to explain just what they did end up with instead struck a few organic chemists as intellectually honest and necessary. Styrene, celluloid, and bakelite were discovered in this accidental manner. (Bakelite was used to make the housings of virtually all early telephones and was one of the first products of a company now known as Union Carbide.)

Third, biochemists were beginning to resolve the structures of many biologically important molecules. To the surprise of many biochemists — but not to the surprise of many bioorganic chemists — many of these molecules were composed of repeating units of very similar protein, fat, or carbohydrate building blocks. Could simpler organic compounds form similar chains? they asked themselves.

Fourth and finally, a small community of scientists, usually spurred on by wartime shortages, deliberately sought to make artificial rubber for tires or artificial silk for parachutes. Successes came first for synthetic rubber. World War I was the catalyst, and the isolation of Germany from the tropical supply of natural rubber was the motivation. Isoprene was identified as the real monomer (building block) of the natural rubber most in use at that time. Soon, a workable isoprene-based synthetic rubber was being produced. However, it was soon discovered that chemically combining alternating units of two simple but different carbon compounds — a process called copolymerization — into long chains gave an even better rubber. This second type of rubber survived the post-war crash of interest in synthetic rubber, because it proved superior to once again abundantly available natural rubbers.

Owing to America's lead in the automotive industry at that time, and the early advances in vulcanization made by Charles Goodyear, America, too, had a substantial interest in rubber-type polymers.

America's entry into synthetic rubbers came from a collaboration of Notre Dame University and DuPont, and led to neoprene rubber. Even today neoprene is used in diver's wet suits.

A lesser-known firm came up with a rubber that is now part of the firm's name: thiokol. Morton-Thiokol, a company tremendously successful in many fields, is now remembered most for its connection with the failure of the synthetic rubber "O" rings in the Challenger rocket booster.

Starches and cellulose proved to be early successes of the biochemical group. They found that they could prepare and even preliminarily analyze ever increasing numbers of them, usually from plant materials. A number were able to be brought to a state of near-crystallization (recall once more that crystallization was considered evidence of purity and stability) and were then able to be analyzed by X-rays. Michael Polyani, a British physicist, suggested that cellulose was a long chain of connected units on the basis of X-ray crystallography, but his findings were quite simply disbelieved. It just seemed too unlikely for "pure" molecules to be so long and come from such a "messy" situation.

Moreover, it seemed unnecessary to many serious university scientists at that time to dwell on just how polymers were constructed. Many useful polymers seemed to be discovered by mere factory engineers and academically unschooled inventors without an understanding of basic organic chemical units. Indeed it was thought academically ruinous to spend too much time on polymers, since it was assumed that they were either a loose association of more basic blocks (in which case it was more elegant to just study the blocks themselves) or just a mess suitable for trade school studies. In fact, the early German epithet for the field was "Grease Monkey Chemistry."

It would take an exceptionally determined Swiss chemist, Hermann Staudinger, to make the field academically respectable. By age thirty-nine he had already made his name in traditional organic chemistry, and could afford to take on this unpopular field without losing his reputation. Much like Liebig, he had strong antipathies for people with whom he disagreed professionally. But also like Liebig, he was generally right. More than any other single scientist, he worked to convince the doubting academic world that his "macromolecules'' were not loose, informal, or messy aggregates. Rather, he argued, they were largely well-ordered molecules with uniformly repeating building blocks throughout the molecule. He argued that even if the polymer molecule did have varying lengths according to the circumstances of its synthesis, it was the wonderful regularity of its internal components that gave a polymer most of its properties, not its length. His principle vehicle for persuasion was the increasingly influential journal *Makromolekulare Chemie*, (''Big Molecule Chemistry'') which he founded and edited.

Staudinger made a number of distinguished converts. Wallace Carothers, who trained at the University of Illinois and then taught at Harvard, was one of the first of these converts in America. He was eventually told by the dean at Harvard that his chosen field of macromolecules was not too promising, and was asked to move on. He then became DuPont's first full-time research director. Carothers had an initial success in collaboration with Fr. Nieuwland of Notre Dame on neoprene rubber, but achieved his most lasting fame as discoverer of an artificial silk, today most commonly called nylon. Unfortunately, Carothers was a chronically depressed and insecure person. Despite his increasing scientific status and secure support from DuPont he committed suicide after he learned of the death of his twin sister. Fortunately, one of his assistants at DuPont picked up the pieces and went on to sustain America's rise to eminence in pure and applied polymer studies.

Paul Flory was Carothers' most important scientific successor. He was one of the first to apply statistical methods to calculate the chemical and physical properties of polymers and develop useful approximations. Flory's success owed much to Staudinger's notion that pursuing the exact size or exact properties of individual polymer molecules — the method used in the rest of chemistry — was not necessary to achieve workable results. He achieved those results by using averages calculated from many samples. Flory went on to influence polymer research at both American industrial giants (DuPont, Exxon, and Goodyear) and prominent scientific schools (Cornell, Carnegie-Mellon, and Stanford.) Flory was not without scientific rivals however, and many major advances to the physics of polymers were made by his principal competitor, Walter Stockmayer of Dartmouth.

Today, polymer science is virtually a freestanding scientific and engineering field, but it retains its roots in organic chemistry. Indeed, Staudinger was to take the Nobel Prize in Chemistry in 1953 and Flory in 1974. Few large organic chemistry departments will be without some polymer chemists and few schools or industrial centers with an interest in polymers will be without some organic chemists.

## SORTING OUT THE JOURNALS

### Journals of General Organic Chemistry

Virtually every national chemistry society that publishes a variety of specialized journals will have one that emphasizes organic chemistry. Nonetheless, it is probable that organic chemistry will dominate that society's nominally general journal as well, so great is its influence. In spite of this generous support, for-profit publishers have fielded other highly successful titles in organic chemistry with advantages of exceptional quality, speed, or historical lineage. (See Figures 1 and 2.)

The leading title by most measures is the American Chemical Society's *Journal of Organic Chemistry*. It is, nonetheless, hard-pressed by the pair *Tetrahedron* and *Tetrahedron Letters*, for-profit entries from Pergamon. This development gives pause for serious consideration on the part of librarians as these two journals are among the most expensive in any collection. Can these titles be worth almost $3,000 each annually?

It is very hard to argue that they are not essential, and indeed, highly distinguished. Both titles were founded by two Nobel Prize winners (Robinson of the U.K. and Woodward of the U.S.) and are currently edited by another (Derek Barton, formerly of the U.K., now at Texas A&M), with the assistance of at least nine other Nobel Prize winners on their various advisory boards. (Indeed, a connection with either journal appears to be a good predictor of future Nobel Prizes as well.) *Tetrahedron Letters* is the original camera-ready copy, rapid communications journal in science — a format that has proved tremendously successful. *Tetrahedron* is a leading source of full-length papers, and offers as bonuses highly cited

FIGURE 1

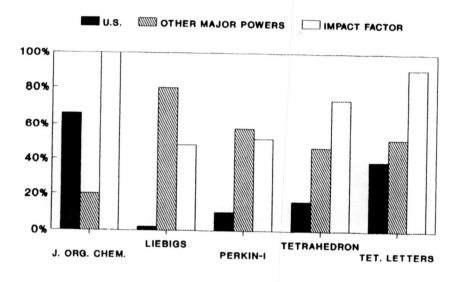

# Contributors and Relative Impact
# Leading Organic Journals

█ U.S.     ▨ OTHER MAJOR POWERS     ☐ IMPACT FACTOR

overviews (called *Tetrahedron* Reports) and special topic theme issues (called *Tetrahedron* Symposia-in-Print). It appears to this author that not even a small undergraduate college could go without
both of these titles and still give its faculty and students a representative exposure to leading developments in this field.

The other two leading journals in Figures 1 and 2 also have distinguished pedigrees. The *Journal of the Chemical Society, Perkin
Transactions I* (Royal Society of Chemistry) features some of the
best synthetic and bioorganic work from Britain and the Commonwealth countries. The journal is, of course, named for the founder
of English organic chemistry, William Perkin. Likewise, *Liebigs
Annalen der Chemie* (VCH), the direct successor of Justus Liebig's
original 1836 journal, is the most important organic chemistry entry
from Germany. While both of these latter titles should be in every
serious collection, *Liebigs* is at a disadvantage. It still publishes

FIGURE 2

# Comparison of Number of Papers, Costs, and U.S. Market Penetration

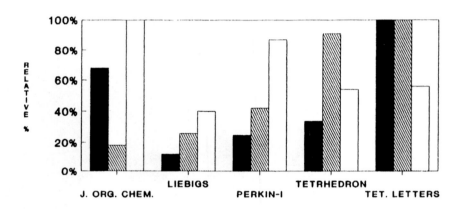

predominantly Central European work, and does so largely in German despite recent changes allowing English. It has failed to attract as many American papers as its historical prestige would otherwise warrant.

## Journals of Synthetic Organic Chemistry

Synthesis remains the glamour field of organic chemistry, and it is not surprising that a number of titles are specifically devoted to it. (See Figures 3 and 4.) All of these titles compete not only with each other, but with important syntheses announced in general chemistry and general organic chemistry titles (particularly the *Journal of the American Chemical Society* in the first instance, and *Tetrahedron Letters* in the second). It is also not surprising to see a leader among the specifically synthetic titles from Germany. This time, however, the journal, *Synthesis* from Thieme, is largely in English and quite

FIGURE 3

# Contributors and Relative Impact
# Leading Synthetic Organic Journals

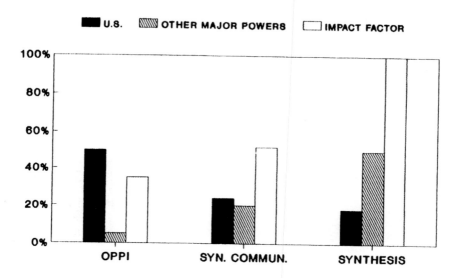

international in contributors. Its closest competitor, *Synthetic Communications* from Dekker, has a higher proportion of papers from the U.S., but publishes fewer and less frequently cited papers overall, and at a higher price. Nonetheless, *Communications* is a sound second choice.

Completing the collection is another American title. *Organic Preparations and Procedures International* (OPPI Inc.) is a fairly slender journal with a high proportion of U.S. papers. Its cost per paper, however, seems steep.

## Journals of Heterocyclic Chemistry

Two highly competitive journals cover most of the world's heterocyclic chemistry. *Heterocycles*, an Elsevier title issued out of Japan, is the leader in number of papers, impact factors, and high price. However, the *Journal of Heterocyclic Chemistry* (Hetero

FIGURE 4

# Comparison of Number of Papers, Costs, and U.S. Market Penetration

Corp.), is clearly the leader in the number of U.S. papers, and has about three-quarters of the impact factor for about half the price. Virtually every sizeable organic chemistry collection should have both titles, with smaller ones preferring the *Journal*. The Russians have a translated title, *Chemistry of Heterocyclic Compounds* (Consultants Bureau for Plenum) that should be in the largest collections as well.

## *Journals of Bioorganic Chemistry*

This small field historically had two journals covering the field. The *Journal of the Chemical Society, Perkin Transactions I*, has placed the most stress of any general organic title on bioorganic chemistry, and has served British and Commonwealth authors particularly well. Nevertheless, *Bioorganic Chemistry* (Academic) is clearly the leader with the rest of the world, and is the first choice for small collections.

This situation may change in the 1990s. The American Chemical Society has announced a new title, *Bioconjugate Chemistry*, that covers parts of bioorganic chemistry. Historically, however, new American society journals tend to shift some American papers out of foreign-based outlets without really damaging their necessity for serious collections. It remains to be seen if American-based *Bioorganic Chemistry* will be seriously affected, particularly since it already has a larger scope.

## Journals of Medicinal Chemistry

The American Chemical Society has long published the leading journal in this field, the *Journal of Medicinal Chemistry*. (See Figures 5 and 6.) For most of the last two decades, its principal competition for work on the chemical manipulation of drugs was from journals of pharmacy that contained some organic synthesis work, most notably the *Journal of Pharmaceutical Science* from the American Pharmaceutical Association.

Since then, however, two foreign journals have been of increasing interest to American medicinal chemists. *Chemical and Pharmaceutical Bulletin* (1953), a journal from the Pharmaceutical Society of Japan, is now entirely in English. It has shown the kind of progress more commonly expected of Japan only in physical science fields. However, care should be exercised in considering the large number of papers it publishes as a decisive factor. A fair number of these papers do not deal with medicinal chemistry topics, although that subject dominates.

Much like the publishers of the Japanese contender, the French Society for Therapeutic Chemistry has abandoned publication in the national language and has transformed its own national journal into a truly *European Journal of Medicinal Chemistry*. This is also quite a dramatic change. The French have been language-conservative and, unlike the West Germans, Dutch, and Swiss, have not had as much success in attracting competitive papers from beyond their borders. But a precedent has been established with the stunning success of a Franco-Italian collaboration in another field, *Europhy-*

FIGURE 5

## Contributors and Relative Impact
## Leading Medical Chemistry Journals

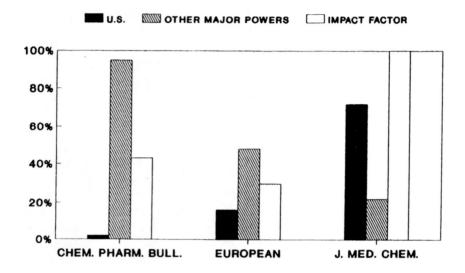

*sics Letters*, and this medicinal chemistry title could well follow suit.

It should be explained that most medicinal chemists will need titles in pharmacology, in addition to the chemically oriented titles just mentioned. This is true even if the American medicinal chemists in your clientele do not publish within those pharmacology titles very often, or rarely do so independently of other health professionals. This is owing to a division of labor in drug design and testing that is stricter in America than in some other countries. In the U.S., at least, medicinal chemists are typically not licensed physicians or pharmacists. These chemists cannot, on their own initiative, conduct human trials of new drugs. Pharmacologists, who may be either licensed physicians or pharmacists, generally have greater authority in this regard. All three specialists — pharmacists, physi-

FIGURE 6

## Comparison of Number of Papers, Costs, and U.S. Market Penetration

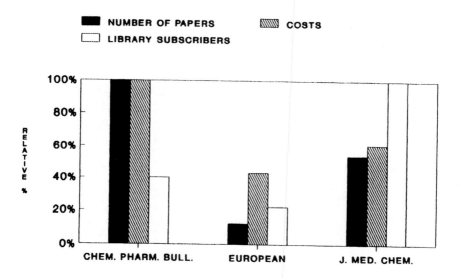

cians, and medicinal chemists — typically work in large teams, with the medicinal chemists responsible primarily for the synthesis and tailoring of the actual medicines using their knowledge of organic chemistry. Test results on humans and even on most experimental animals are usually published by the MD's or D.Pharms in leading pharmacology journals. For U.S. authors, these are the *Journal of Pharmacology and Experimental Therapeutics* or *Molecular Pharmacology* (both from Williams and Wilkins). European work is most prominently covered in the *British Journal of Pharmacology* (Pergamon), the *European Journal of Pharmacology* (Elsevier), and *Naunyn-Schmiedebergs Archives of Pharmacology* (Springer). A more detailed analysis of these titles will be presented in a chapter on physiology and pharmacology in the author's companion work, *Making Sense of Journals in the Life Sciences*.

### Journals for Natural Products Work

As Figures 7 and 8 indicate, a fairly small number of journals cover natural products chemistry. While all four titles in the figures are linked by overall mission, two are more exclusively dedicated to medicinal chemistry concerns. There are geographic patterns that might influence library selection as well. Since 1938, the *Journal of Natural Products*, from the American Society for Pharmacognosy (pharmacognosy is the discovery of drugs from natural sources), has served the needs primarily of U.S. authors. *Planta Medica*, from Thieme (for the German equivalent of the American society just mentioned), does much the same for Continental European authors and, today, does so mostly in English.

The British effort (and much of the Commonwealth effort) is more diffused and involves at least three journal types. First, British authors place a substantial number of natural products papers into

FIGURE 7

# Contributors and Relative Impact
# Leading Natural Products Journals

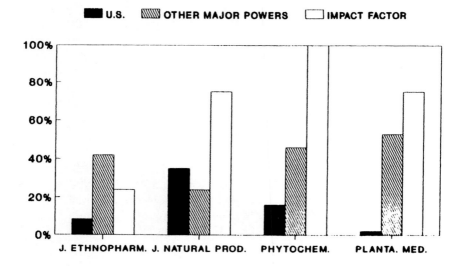

FIGURE 8

## Comparison of Number of Papers, Costs, and U.S. Market Penetration

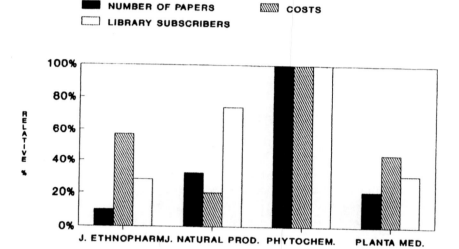

■ NUMBER OF PAPERS     ▨ COSTS
☐ LIBRARY SUBSCRIBERS

general organic journals, particularly both *Tetrahedron* versions and *Perkin I*. Second, they have one of the best plant biochemistry journals in *Phytochemistry* from Pergamon. While *Phytochemistry* is not exclusively devoted to plants of medicinal potential, it covers much of that field. Third, the Royal Society of Chemistry publishes the best review in the field: *Natural Products Reports*.

The final title in our discussion may be the most fascinating for the general reader: the *Journal of Ethnopharmacology* from Elsevier. While the level and orientation of most papers is highly technical, papers detailing the use of drugs in primitive cultures as reported by explorers and anthropologists are occasionally accepted, and similar colorful details are often included in regular laboratory papers as well. An added feature for the scientist is recurring abstracts of papers appearing in other journals of interest to ethnopharmacologists.

The *Journal of Natural Products* is the best choice for small collections, followed closely by *Planta Medica*, based on its sharper focus on clinical potential and more tolerable total cost when compared to the other remaining titles. But even moderate collections should have *Phytochemistry*, given the scientific and social importance of drug research, and the potential use of this journal for plant scientists from other academic departments. Given its high cost per paper, The *Journal of Ethnopharmacology* must be considered an option mostly for the very largest collections. Even schools with an anthropology department might give pause before sharing their budget for a journal with this high price tag.

### Journals for Physical Organic Chemistry

Two titles dominate this area. *The Journal of the Chemical Society, Perkin Transactions II* is the well-established leader. It has been challenged since late 1988 by Wiley's *Journal of Physical Organic Chemistry*. The challenger will bring on two changes. First, a higher proportion of American work will shift to the challenger. Second, the new title will shift more of this work away from journals of general physical chemistry—journals within which organic chemists have never been particularly comfortable. It has been argued that in general physical chemistry journals, the author, out of a fundamental interest in physical phenomena, chooses the chemical system that best illustrates his or her idea. If the paper helps an organic chemist, reasons the physical chemist, this is fine, but not the motivation of the paper in the first place. In the thinking of straightforward physical chemists, the ripeness or maturity of the idea is what determines when and if the idea should be published. By contrast, in physical organic chemistry journals, the author, usually an organic chemist at heart, understands the needs of colleagues in organic synthesis or natural products isolations. He or she preferentially explores practical physical problems that actually crop up in organic chemistry, and offers organic colleagues a physical explanation around some difficulties or unexpected results. Both the new and established physical organic titles should be in most organic chemistry collections.

## Journals of Polymer Chemistry

Given the enormous practical need for plastics worldwide, it is not surprising that there are dozens of polymer journals. One multiplying factor is nationalities, with each national society having a polymer journal. The other factor is the "sectioning" of an originally larger polymer journal into separate journals. These sections are divided up along component disciplines within polymer science or by special journal features.

It should not be surprising then to find that the *Journal of Polymer Science A - Chemistry Edition* also has a *B - Physics Edition* and a *"Letters" Edition*. Likewise, the *Journal of Macromolecular Science - Chemistry* section also has a - *Physics* section and a section called - *Reviews in Macromolecular Chemistry and Physics*. Most of these partitioned journals have an "Applied" version, and occasionally a "Biochemical" version as well. (Recall that many proteins, fats, and carbohydrates are essentially long strings of similar basic units linked polymer-style.)

This discussion will focus primarily on the polymer chemistry sections. These versions maintain the closest relation to their organic chemistry roots, and have the merit of being needed in both many organic and polymer chemistry departments. (See Figures 9 and 10.) The leader in this field, *Macromolecules*, comes from the American Chemical Society. Not only is this title the most highly cited, it publishes the most papers and at the lowest per-paper cost. The second place title is hotly contested. *Makromolekulare Chemie* (Huthig) is historically very important—it was Staudinger's principal journal and it is the next most frequently cited per paper. However, the *Journal of Polymer Science - Polymer Chemistry Edition* from Wiley has a higher proportion of U.S. papers and at a more reasonable cost. It should get the nod in financially strapped collections where both journals could not be taken. The *Journal of Macromolecular Science. Part A. Chemistry* from Dekker rounds out the basic chemistry collection.

Departments of chemical engineering and freestanding polymer science institutes should get all the "physics" and "applied" sections of these titles and consider certain relatively "unsectioned" polymer titles. Three that are notable include Butterworth's *Poly-*

*mer*, Springer's *Polymer Bulletin*, and Pergamon's *European Polymer Journal*. *Polymer Communications* from Butterworth is the best letters journal.

Studies of natural and synthetic rubber remain important and are best covered by the American Chemical Society's *Rubber Chemistry and Technology*. Biochemistry departments would be well advised to get Wiley's *Biopolymers* and Butterworth's *International Journal of Biological Macromolecules*.

Reviews in macromolecules apart from Dekker's entry mentioned above, are featured in Springer's *Advances in Polymer Science* and Pergamon's *Progress in Polymer Science*.

FIGURE 9

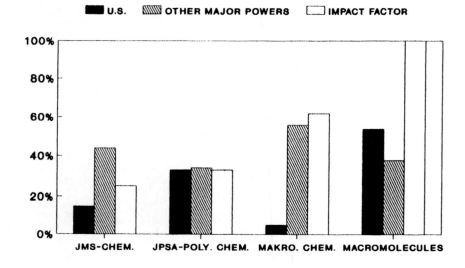

**Contributors and Relative Impact
Leading Polymer Chemistry Journals**

FIGURE 10

# Comparison of Number of Papers, Costs, and U.S. Market Penetration

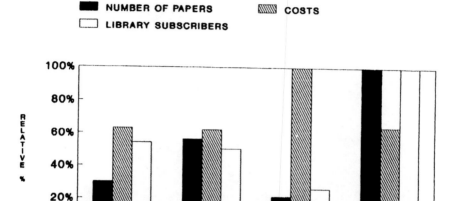

# Chapter 4

# Physical Chemistry, Chemical Physics, and Their Journals

## BACKGROUND

In a banquet speech honoring a Nobel Prize-winning physical chemist, the speaker explained the chronic dilemma of physical chemistry and chemical physics: "These new theories suffered from the misfortune that nobody really knew where to place them. Chemists did not recognize them as chemistry; nor physicists as physics." Librarians today are faced with a related dilemma: which departmental collection gets these journals? The problem is not one of just chemistry vs. physics; today, libraries serving engineers, mathematicians, and computer scientists are just as likely to need to need some of the journals discussed in this chapter.

### All-Stars Who Didn't Get to Play as a Team

The history of physical chemistry and chemical physics is dominated by a series of isolated heroes. The isolations were sometimes chronological: an idea would be advanced by a scholar of one generation, but then ignored for two or three generations before another scholar realized its importance and developed it further or applied it. The isolation could be geographical: a development could be pursued in one, typically less developed country for some time without being noticed in another country with a larger and more internationally influential scientific establishment. Finally, the isolation was one of disciplines and occupational priorities: a concept was pursued by an engineer for one reason and then not noticed by mathematicians, chemists, or physicists because each discipline had a different angle on the problem or a lower place for this problem on its own agenda.

### *The Working Engineer's Experiences with Temperature, Pressure, and Work Efficiency Come Together to Form the Science of Thermodynamics*

One of the most ironic twists in physical chemistry is that today's largely molecule-level specialty of thermodynamics had its roots in bulky, inefficient steam engines and bubbling vats of brew.

The early 1700s in Britain had seen a number of steam engines that were principally used to pump water out of coal mines. Many tended to use up almost as much coal as they as helped to mine. In the 1770s, however, James Watt, a largely self-taught Scottish inventor, made an improvement that actually got an extra 5% to 7% worth of work out of these boilers. This small gain, and certain mechanical improvements that went along with it, helped to make the steam engine a prime force of the early industrial revolution. Subsequent gains in steam energy efficiency were very slow in coming, however, and this lack of of progress could not be explained well.

Not until the 1820s did a mathematically well-trained engineer, Sadi Carnot, analyze the situation correctly. Carnot was a member of an especially industrious family that was held back in the years before the French Revolution by their lack of royal blood. (The family was also characterized by reusing the same given names over and over so that tracing distinctions among the Carnots often requires knowing their Confirmation names!) The only branch of the officer corps open to commoners like the Carnot family was military engineering. However, this branch did have the advantage of providing Carnot the father, Carnot the son, and Carnot the cousin with a free, high quality scientific education. Largely through his application of mathematical analysis, Carnot the father developed into the modern world's first military efficiency expert. He was to achieve fame for the systematic transport of supplies during the eras of the French Revolution and Napoleon. In his lifetime he was officially proclaimed "The Organizer of Victory." (Today he'd be called an operations researcher and have an endowed chair in a business school.) Carnot the cousin became a successful politician after his successful military career. (Carnot the cousin's own son became the President of France.) This left the quietest of the Carnots, Sadi the son.

Here was a young man desperately wishing for recognition as a real scientist. Throughout his sad and brief life — he died at age thirty-six — Carnot entered his mathematical manuscripts to scientific essay contests. Sadly for Carnot, these papers won only a string of "Honorable Mentions" in his own lifetime. Historically, however, one of these modestly judged efforts paid off for the science of thermodynamics. In what amounted to a small privately printed booklet, Carnot closely tied together the notions of "force-generated" and "heat-given-off" by machines. He showed that a certain inefficiency involving wasted heat was inevitable in virtually any boiler or steam engine. By analogy, he suggested that this limited efficiency relationship of heat and force might apply to virtually all work-accomplishing situations. The strength of his arguments lay in an adept use of calculus and picturesque French language. But Carnot's approach was largely beyond the understanding of Watt's successors. They were practical men who reckoned from steam tables and diagrams rather than from formulas, and were not typically fluent in French.

It was at least ten years before a successful railroad engineer, Benoit Clapeyron, advanced thermodynamics at all, and that was almost an accident. Carnot's work became better known because Clapeyron's French-language book on railroad engineering was translated into both English and German, indirectly translating the extensive passages of Carnot quoted by Clapeyron as well. Clapeyron adapted some of Carnot's expressions and formulas into steam tables enabling them to be better understood by working railroad engineers.

Another working man, a Scottish brewer named James Joule, had a more immediate assist in his thermodynamics discovery. He was attracted to science through the popular lectures of John Dalton, a hero of the chapter on inorganic chemistry. Self-educated, but amazingly observant and handy in making thermometers, Joule maintained a peculiar diary detailing how virtually any motion in fluids involved shifts of temperature and therefore of energy balance. Joule's work showed that the extremes of steam heat were not necessary to observe certain thermodynamic laws. But the unschooled Joule could not, at first, readily derive standardized equations that could be evaluated by established scientists. Fortunately another Scot, Professor William Thomson, was a customer of Joule's and a mathematical genius attached to the University of Glasgow. He befriended Joule, derived those equations, and, among other achievements, demonstrated that certain thermodynamic effects were operating consistently even at very cold temperatures. Thomson was made a Baron for his achievements, and chose the title Lord Kelvin, a name now recalled in America for "Kelvinator" refrigerators. Thomson was one of the first to write a text specifically devoted to thermodynamics, and included the work of Carnot, Clapeyron, and Joule alongside his own efforts. (With all this attention and encouragement, Joule completed his schooling and became a first-class physicist, by the way.) With Thomson's book, a picture was slowly emerging of laws of energy transfer which operated across a wide range of temperatures and conditions. A question remained: could such laws come into play in chemistry?

### Finding Thermodynamics in Chemistry and Discovering Analytical Thermal Chemistry Along the Way

That question was answered by at least two scientists who, for decades, had no large audience. The first to suggest an answer in the affirmative was Germain Hess, a Swiss who lived and worked largely in Russia. He was a medical doctor by training, but for a number of years also worked on the side as a minerals consultant, owing to his posting in a remote mining district in Siberia. For years, he performed chemical analyses of various minerals and ores. When he was posted back to the major cities of Russia, he began to compare his chemical reactions with previously reported chemical literature on alternative ways to achieve the same products. He paid particular attention to the changes in temperature generated by these reactions, or the amounts of heat required to force them along. He discovered, much to his surprise, that barring a catalyst, the total energy required to form a given chemical product was about the same, no matter whether the steps on the way to that product were many or few, or the routes somewhat different. This suggested that the energy required to change certain starting materials into a new compound was a thermodynamic characteristic or constant of that new compound. It occurred to Hess that assembling such data would be very useful for the identification and handling of compounds. Hess didn't realize it, but just such an activity has resulted in the rise of the thermal analysis portion of thermodynamics. Hess' announcement would have been revolutionary if it had had a better hearing at the time. Alas, Hess' initial connection to the outside chemical community was largely with the aging and formerly-feared chemical critic, Jakob Berzelius of Sweden. With the unpopular Berzelius going into decline, Hess had no international champion. Hess' first champion of any sort was just as isolated as Hess. He came from the scientific wilderness of America!

Professor Josiah Gibbs of Yale was that champion, and one of the most quietly diligent scientific literature analysts of his time. He devoted many years to reading the reports of engineers, mathematicians, and chemists. Gibbs worked up extremely long and mathematically sophisticated articles conclusively demonstrating that thermodynamic principles operated in virtually all chemical phenomena. He showed how energy shifted when a compound shifted from a solid phase to a liquid phase, and then to a gas phase. He developed the notion of an intricate balance between potential energy and freed energy. However, these meticulous 100-page-long blockbusters appeared in the *Transactions of the Connecticut Academy of Sciences*, a journal of no standing in scientifically dominant Europe. It was over twenty years before translations of Gibbs work — which included mentions of Hess' hitherto ignored work — appeared in academically acceptable German-language periodicals.

## The German University Develops an Interest that Becomes Thermophysics

The prestigious world of German, Austrian, and Swiss academic physics did not get involved in thermodynamics through the same avenues as engineers or chemists. Physicists did not place much emphasis on studies involving chemical reactions in liquids. They used the analogy of the steam engine only in that steam was a form of gas. Unfortunately for thermodynamics, early physicists did not place a great importance within their own careers on the thermodynamic behaviors they did observe. Some of the most important thermodynamic discoverers became much better known for other discoveries they made. Hermann Helmholtz, for example, was a famous German physician-turned-physicist. He developed the opthalmascope for examining the eye, and various devices for measuring hearing and nerve impulses. While his pronouncements early in his medical career on the relationship of work and heat would later be shown to be fundamentally correct, Helmholtz could not get them accepted into physics journals because his medical credentials made them seem suspect. He had to print them up privately in pamphlets that he proceeded to give away free. Only when he gave up teaching medicine for teaching physics some twenty years later was this thermodynamic work taken seriously. Then it was thought to be highly novel and exciting. Indeed, for the German-speaking world, Helmholtz is the founder of thermodynamics.

By contrast, Rudolph Clausius went directly into physics, and quickly devoted himself to thermodynamics. He offered a more mathematical treatment of thermodynamics than anyone else (save the still-unrecognized Gibbs). Clausius developed the thermodynamically key term "entropy," leading to a second law of thermodynamics. Clausius was, however, extremely pompous. He viewed himself as a spiritual descendant of the ancient Greek physicist-philosophers and had rather severe ideas about the purity and ultimate truth of his message as compared with the work of others. This led him to argue even with those who agreed with him substantially, but not unreservedly. Clausius' concept of entropy was initially controversial. When it appeared to be accurate, the controversy shifted towards who should get credit for the law arising from it. Clausius' chief rival was the British scientist Tait, who pointed to the work of other British scientists such as Joule and Kelvin as backing the Tait claim. Clausius responded by citing Continental antecedents as far back as Carnot (the son) and as recent as Helmholtz. The nationalism involved detracted seriously from attention to thermodynamics itself.

## Success in the Thermophysics of Gases Reinvigorates Interest in Liquid Phase Chemical Thermodynamics

Curiously, it was the leading British physicist of the age, James Clerk Maxwell—a man far better noted today for his electromagnetic theory—who resolved the Clausius-Tait dispute shortly before his own death. Earlier in his distinguished career, Maxwell had pondered the interplay of gases and heat in a widely used textbook, which was itself a revision of Lord Kelvin's early thermodynamics book. Maxwell proved that it was the collisions of gas phase molecules against their containers that registered as the phenomenon of vapor pressure. He suggested that increasing the heat of the container increased the energy of those collisions against the container, causing the pressure to rise. (The interplay of heat and pressure in gases had long been observed by chemists, but many of them still held that "heat" was of itself an ether-like substance, not something that energized the molecules.)

In light of the entropy controversy of the time, Maxwell looked over his old book on thermodynamics, revised it in line with Clausius, and reissued it with an introduction that spread credit around, but most generously so for Clausius. (Ironically, Clausius was so embittered by the controversy that he failed to grasp how certain assumptions and figures of speech used by Maxwell actually strengthened his own theory.) Shortly thereafter, Maxwell was one of the first British scientists to note the work of Gibbs. While he managed to send letters endorsing Gibbs to a number of prominent Continental physicists, Maxwell died before yet another revision of his important textbook acknowledging Gibbs could be prepared. But Maxwell's letters had done the trick. The Germans began reading the work of Gibbs. With Gibbs finally coming to the fore, liquid phase work such as that of Hess, as cited by Gibbs, also got an important second look in the chemical community.

The molecular excitation view in Maxwell's thermodynamics was greatly extended by Austrian physicist Ludwig Boltzmann. Boltzmann worked out the mathematics to an even higher degree, and a number of rules and constants in the thermodynamics of gases today bear Boltzmann's name. Boltzmann's principal contribution to physical chemistry, however, may have been his encouragement of a young physicist named Walther Nernst to put his physics background to use in the liquid world of chemistry. The notion of an interdisciplinary career was highly unusual in that era, and was often regarded as a defection from the ranks of the truly committed disciplinary scientist. (Recall the story of von Helmholtz!) Nernst may also have been encouraged by the example of Fritz Haber, a man who went from chemistry to chemical engineering, a more modest and acceptable transfer. Haber was the first chemist to make extensive use of the Maxwell-Boltzmann concept of gases by developing a large scale operation to extract nitrogen from the atmosphere. This feat at the turn of the century eventually won Haber the Nobel Prize in 1918. Boltzmann did not live long enough to see the success of his gas laws, or the outstanding success of his influence on Haber or Nernst. Depressed that neither his work nor advice was taken seriously, he committed suicide in 1906.

### Electricity and Water Spark the Development
### of Electrochemistry and Solution Chemistry

Electricity in one form or another has served to amuse lay people and puzzle scientists since classical times. The bulk of electrical research in the 1700s and early 1800s tended towards demonstrations of curious phenomena: that electricity was present in lightening (Benjamin Franklin), that electricity could cause muscles to twitch (Galvani), that stacks of coins of at least two differing metals in alternating layers could make a primitive battery (Volta), and so on. For our purposes in physical chemistry, electricity would make its mark as soon as it became involved with wet chemicals in a way that sounds dangerous now, but was reasonably safe then given the low voltages generated.

### Faraday and His Times

Probably the first significant electrochemist was Humphry Davy, a wealthy British gentleman scientist. He hit upon the idea that stacks of metal plates or coins worked even better as electrical batteries when placed in an acidic solution. (Your car battery is very much like this, almost 200 years later.) Davy was one of the first chemists to comprehend the electrochemical breaking apart of compounds into their component atoms. He dissolved many compounds formerly thought to be insoluble in a water bath in which he had already positioned his battery-charged electrodes. The aging Berzelius of Sweden tremendously extended Davy's work, and it is largely for thousands of these systematically cataloged, electrochemically driven breakdowns that the domineering chemical critic from Sweden is most fondly remembered today. Both Davy and Berzelius ventured the idea that the atoms, newly liberated from their compounds and then attracted to either the positive or negative electrodes, were telling us something about their own positive or negative chemical charge, a finding that turns out to be essentially correct.

Davy was the first principal scientist of the Royal Institution in London. The Royal Institution was essentially a gentleman's scientific society with a bit more of a socially uplifting mission than matching agencies of the nobility in archrival Paris. As Davy's workload increased and his health deteriorated, he sought out an inexpensive assistant who would also serve as his wife's servant. A poor young man named Michael Faraday, who studied chemistry at night, had long been an admirer of the aristocratic Davy. Faraday was an apprentice binder and took books and journals home with him after hours. On his own initiative, Faraday transcribed, illustrated, and bound the public lectures of Davy, and sent them to Davy as his ploy to get an introduction to this celebrity scientist.

As it turned out, this servant boy, who even then caused librarians a headache through bindery delays, would become England's greatest scientist in the era between Dalton (1766-1844) and James Clerk Maxwell (1831-1879). Faraday succeeded his boss after Davy's chemicals and personal lifestyle got the best of him. (Davy lived very well, and additionally, developed the very bad habit of drinking his chemicals to record their taste and effects: a Dr. Jekyll and Mr. Hyde approach.)

Faraday was deeply religious and always conscious of his extremely modest social origins. He made the Royal Institution a far more democratic and humane organization, and was very much like Dalton of the generation before him in promoting popular lectures. Charles Dickens, the great socially concerned novelist of the 1800s, was one of Faraday's most ardent supporters. Faraday's principal claims to overall scientific fame were largely in the area of electricity, and in Britain he holds the status that Americans of a later generation would grant to Thomas Edison. In terms of physical chemistry, his greatest contributions were in studying the effects of increasing or decreasing the electrical current in his electrochemical tanks, and varying the concentration of the materials to be broken down. The ability to precisely manipulate these factors became extremely important for the next generation's study of materials dissolved in solutions.

### Electrochemistry Arcs in Northern Europe

Advances in electricity made it possible for Friedrich Kohlrausch, a German chemist of the next generation, to use alternating current to study solutions. In particular, he was able to work with much more dilute solutions than did Faraday, and managed to go much higher and much lower in the amounts of electrical current employed. Kohlrausch discovered that at given settings, specific dissolved elements migrate to the electrodes. By altering the settings, given elements could be detected at their distinctive settings. Through his finding, Kohlrausch became the father of the analytical chemistry side of electrochemistry. (See the chapter on analytical chemistry). To a substantial degree, chemical electrodes such as those used to test water today work very much the Kohlrausch way.

This ability to work with very weak solutions and, after Kohlrausch, to work at similarly weak currents, led to the work of Svante Arrhenius, the most important Swedish chemist since Berzelius. Arrhenius questioned whether or not the electrical current that everyone had assumed caused all of the breakdown of the compounds was, in fact, wholly responsible. Could not some of these compounds break down entirely on their own without the electrical current? As it turns out, not even his PhD dissertation committee believed that Arrhenius was right. They awarded him a "D," enough to pass, but too low a grade to get a good job in his native Sweden. This was ironically fortunate, for Arrhenius then sent copies of his thesis all over Europe, looking for a job offer from anyone who could agree with its findings. He got nibbles from only two scientists, Jacob van't Hoff, a Dutchman, and Wilhelm Ostwald, a well-established German chemist. While he worked amicably and briefly with van't Hoff, his most important encouragement came from Ostwald.

Ostwald had a personal antipathy for what he considered to be Germany's overemphasis on organic chemistry. Liebig's astonishing success as champion of organic chemistry two generations earlier had led to an overabundance of young postdoctoral students in organic chemistry and the crowding out of nonorganic papers in journals of general chemistry. Together with van't Hoff, Ostwald founded the *Zeitschrift fuer Physikalische Chemie*, the first journal devoted to the new specialty. It was one of the first places to translate the previously obscure work of that thermodynamicist from the wilds of America, Gibbs. It had done so as a result of the famed letter from Maxwell.

Ostwald could afford to go against the grain because of his early successes in the thermodynamics of liquids. Findings from workers like Ostwald had already paid off handsomely for the German chemical industry. Ostwald, for example, could calculate the point at which a reaction would come to an equilibrium that was no longer productive. He advised manufacturers that changing the conditions of equilibrium subtly—for example, by siphoning off a small portion of the product continuously—often yielded more chemical product eventually than letting the chemical batch go its own way to a natural dead stop. Likewise Ostwald recognized that the final volume of a batch of reacting chemicals was not entirely predictable by adding the individual volumes of the chemicals before mixing. Many reactions had increased volume or released a great deal of heat. Ostwald's advice avoided the boil-over of reaction vats and the rupture of pipelines. In purely scientific terms, Ostwald's generation instilled the idea that the mixing, concentration, electrical conductance, expansion, and heat releasing behavior of chemical solutions were of themselves well worthy of attention. Ostwald was effectively the founder of what later came to be known as "solution chemistry."

After his own fruitful collaboration with Arrhenius, Ostwald placed the young Swede with Kohlrausch. After this third postdoctoral training period, Arrhenius was asked by Ostwald to publish his increasingly confident theory of the natural breakdown of some ionic compounds in the *Zeitschrift*. While the theory was still not consistently true in all situations, it won for Arrhenius the 1903 Nobel Prize in Chemistry. Ironically, Ostwald, his principal sponsor, was to wait until 1909 for his laureate. The man who finally resolved the remaining difficulties in the work of Arrhenius was Dutchman Peter Debye. Debye's papers in the British journal *Transactions of the Faraday Society* were to win the 1936 Nobel Prize.

### Nernst Brings Germany to the Pinnacle of Physical Chemistry Only to See It Fall to Nazi Interference

These were by no means the only Nobel Prizes related to Ostwald and Arrhenius. Arrhenius came upon the young physicist Walther Nernst, the fellow whom Boltzmann had already encouraged to get into physical chemistry. Arrhenius got Nernst a job with Ostwald. Working from the electrochemistry of solutions, and having begun his experience fully cognizant of Gibbs mathematics, Nernst worked up yet another thermodynamic law to win the 1920 Nobel Prize.

Nernst was spectacularly successful as an organizer of graduate programs, scientific conferences, and research institutes. He succeeded Ostwald as editor of the *Zeitschrift fuer Physikalische Chemie*. He was in almost every way the early twentieth century's champion of physical chemistry, in the manner of Liebig's nineteenth century championship of organic chemistry.

Nernst was a short, fat man who invariably sported an old-fashioned pince-nez and fat cigar. A man of deliberately comic pomposity (he performed parodies of himself at social functions), he made an enormous amount of money from consulting and inventions. He used his money and influence to protect those scientists persecuted by the Nazis. When that was no longer possible, he eased their transitions out of Germany, by providing funds and using his prestige to obtain jobs for them abroad.

Nazism devastated physical chemistry in Germany and enriched it in North America. It was an example of the stupidity of Nazi thinking that Fritz Haber, the Nobel Prize winner who tirelessly devoted himself to the German munitions industry during World War I, was forced to leave Germany because he was Jewish. Likewise, Peter Debye, a Nobel Prize-winning Catholic who strongly objected to Nazi demands that he purge Jews at his research institute, was found a job in America at Cornell University by Nernst.

Cornell was to play an especially important role in the early years of physical chemistry in America. Cornell University was specially receptive to Debye because two of its first physical chemists, Wilder Bancroft and Joseph Trevor, had studied in Europe under Ostwald. Indeed Bancroft and Trevor set up the *Journal of Physical Chemistry* along the lines of the *Zeitschrift* . . .

Ultimately, the Nazis silenced Nernst, despite his having won the Iron Cross in the first world war and having lost both of his sons in the German army. His "unforgiveable offense" was allowing his daughters to marry Jews.

## The Shape of Things to Come:
## Modeling Molecules in Three Dimensions Gets Its Start

Ostwald had a partner in the founding of the *Zeitschrift fuer Physikalische Chemie*, the Dutch chemist Jacob van't Hoff. Van't Hoff's career was cut short during a tuberculosis epidemic, but he quite literally added a "new third dimension" to chemical thinking before he died. In his major work, van't Hoff was influenced by Kekulé, the discoverer of ring systems in organic chemistry, and Le Bel, a member of the French dynasty in organic chemistry in the era of Dumas. Van't Hoff published a twelve-page pamphlet in Dutch explaining differences in the chemical behavior of compounds that clearly had the same atoms but somehow behaved differently. He suggested that the differences were ones of the three-dimensional arrangement of the component atoms in space. While Ostwald was quick to recognize the importance of the pamphlet, few other chemists paid much attention. Indeed, it took three years for it to be translated into German, and ten more before it made it into French. The critical reception was lukewarm for two additional reasons: van't Hoff had derived his explanations without laboratory experiments as proof (he argued from proofs in geometry), and he held a modest position at a veterinary school at the time of publication.

Van't Hoff had to set aside his spatial work for a time until he built up a reputation in more traditional chemistry. He made a number of contributions to the thermodynamics of systems in which liquids, solids, and gases were all mixed, and did a good deal of practical work relating to marine mineral deposits. Had not his three dimensional effort paid out, his thorough explanation of chemical "osmosis" would have secured him fame as a solution chemist. As it turned out, van't Hoff's imaginative thinking would be vindicated, but it would also take a new generation of highly imaginative physicists — and chemists who were paying attention to those physicists — to carry the field further.

### Quantum Physicists Visualize the World of Small Atoms

The full exploitation of van't Hoff's themes required from two groups of physicists and one group of chemists a convergence of explorations in visualization. Quantum physicists explored from the small and simple towards the large and complex. The chemists explored from the large and complex downward to the small and simple. Both groups drew data and checked the accuracy of their visualizations with the aid of a second group of physicists, the spectroscopists.

Two kinds of physicists then, approached the problem of the visualization of atoms and molecules. The first group we will discuss, the quantum theorists, was not really the first to arrive on the scene, but their work made sense of findings of the older group, the spectroscopists.

This first, quantum mechanical group was highly mathematical. Essentially, it built mental models that fit certain equations. If you can recall your own high school days, the earliest visualization for the simplest atoms was the "solar-system" model proposed by Bohr. In this system the electrons — atomic particles which chemists focus on as the primary agents of chemical bonding — travel something of a spherical path. There are several layers of these paths, (or shells-within-shells) depending on the number of electrons and their energy. It was soon realized that the spherical model was probably too simple and the shapes of the electron shells altered as more complicated atoms were imagined or as individual atoms were bonded into multi-atom molecules. Nonetheless, physicists insisted on a step-by-step approach. They tended to try to solve the key equations for the basic parts of the simplest atoms before moving on. For the electrons on the outside of an atom, for example, the key equation was the Schrodinger, named after an Austrian physicist. In contrast to chemists, however, quantum physicists were just as concerned about the nucleus as about electron pathways. They ideally hoped for complete consistencies of all parts and particles in a neat mathematical model.

It was to take two innovators to get the field moving in a way that helped chemical physics. An English mathematical physicist named Paul Dirac broke through the logjam concerning the electrons in the simplest atomic model. He was then able to calculate the electron configuration for even the most complicated single atoms. Soon after, an MIT physicist named John Slater developed means of calculating for electrons in many-atom models. While both Dirac and Slater were to gain far more fame for work outside of chemical physics, both were to once again play important roles in the field at the end of their careers.

## Chemists Visualize the Larger World of Complex Molecules

Chemists were less finicky. They needed visualizations that helped them to predict the shape of the electron orbitals in the larger compounds that they dealt with every day. They tended to settle for useful rules of thumb, often without mathematical sophistication, that worked well, if not perfectly. Not only did they tend to discount the nucleus, they tended to discount electron shells that did not actively participate in chemical reactions. They stressed the outer electron layers almost exclusively and ignored the nucleus.

Their first big success came from Ernest Huckel, a Dutch partner of Debye. In the 1930s, Huckel proposed a successful molecular orbital model invariably referred to today as HMO (Huckel Molecular Orbitals) Theory. It required little more than algebra to be applied successfully, much to the surprise of the elaborately calculating physicists.

Another refugee from Nazi Germany, Roald Hoffmann, settled at Cornell in the 1960s, and teamed up with the greatest organic chemist of the time, Woodward of Harvard. Together they wrote up an influential and, once again, disarmingly simple account of the role of symmetry in these bonding orbitals. Eventually, Hoffmann was to refine these concepts further in what he modestly refers to as Extended Huckel Molecular Orbital Theory. In 1981 the Nobel Prize committee gave him the award for these "improvements." He shared that prize with a Japanese chemist, Kenichi Fukui. Fukui actually had a head start on orbital symmetry studies, but he initially published only in Japanese. His approach gained a wider audience only when he switched his principal outlets from Japanese journals to the English-language *Journal of Chemical Physics*.

## *Spectroscopists Provide Evidence that Helps Small and Large Scale Visualizations to Converge*

A major problem remained in that no quantum mechanical physicist, nor any molecular orbital chemist, had ever seen an atom or molecule, or its electron shells or orbitals. Both groups had gone on trusting in something other than direct evidence: advanced mathematics in the case of the physicists, and practical experience in predicting successful reactions in the case of the chemists. But as time went on both groups were to test their approaches against the best available evidence: spectroscopy. It is important to understand a little of the history of spectroscopy in order to understand how it came to play a benchmark role that continues today in chemical physics.

Spectroscopy, as mentioned earlier in the chapter on analytical chemistry, is the interaction of some form of light energy with matter. More specifically, we now know that it is the reactions of light waves and electron shells. Spectroscopy was long considered a curiosity. Recall the famous story (second in importance only to the falling apple) of Isaac Newton breaking up visible light with a prism and then suggesting that white light is essentially a mixture of all the primary colors. Early experiments with materials to find their color-yielding behavior usually engulfed them with flame (to get the sun-like heat) and then channeled the light through a prism or specially calibrated grating to get a spectrum. Most substances did not give off a continuous rainbow. Rather, only certain characteristic lines and only certain colors came through. The assortment of colors and the spacing of those lines they came through is unique to the substance and serves as its "fingerprint." The reliability and scientific interest in spectroscopy in modern times is due largely to two German scientists, Bunsen and Kirchhoff. Together, they had the right combination of chemistry and physics skills to make a chemical physics advance respected in both chemical and physics communities.

The invention of a clean-burning heat source, utterly necessary for sharp-lined spectra, came from Robert Bunsen of Bunsen burner fame. Indeed, Bunsen shares with Fresenius (of analytical chemistry) the title of the greatest chemical instrument maker of the 1800s. The irony of the invention of the modern spectroscope is that early versions required excellent eyesight. It turns out that Bunsen had blown out one of his eyes earlier in a lab explosion. (He was making one of the first organometallic compounds, cacodyl, and was to write up an amazingly understated cautionary note in the manuscript afterwards.) The irony of this situation was similar to that of Beethoven's deafness and his later music!

The mathematical sophistication and theoretical insight in this team came from Gustav Kirchhoff. Kirchhoff's detailed analysis of the lines led to the discovery of hitherto unexpected substances: the elements cesium and rubidium. The idea that the lines also told the scientist something about the electron shells came from a student of Kirchhoff's, Max Planck. Max Planck became the father of quantum mechanics, and his story will be more fully recounted in the chapter on atomic, nuclear, and particle physics.

It was Gerhard Herzberg, a doctoral student at Planck's school, who was to most assiduously exploit Planck's ideas about spectral lines. Herzberg, through fifty years of obtaining and analyzing meticulously detailed spectra, sought to teach both quantum mechanics physicists and molecular orbitals chemists something about the correctness of their separate visualizations. For physicists, his work in small atoms confirmed the fact that the spacing of the spectral lines contained a great deal of information about the location and energy levels of the electrons in their shells. Herzberg's results often coincided with the highly mathematical methods of the early quantum theorists like Schrodinger and Dirac. For chemists, he was to show through his spectra of large molecules that the clash of the electron orbitals of the individual atoms contributing to a multiatom molecule led to deformed or altered orbitals. Herzberg's spectra clearly showed hybrid orbitals, such as those postulated by the rule-of-thumb theorists. This confirmed the ideas of Slater, Huckel, Woodward, Hoffmann, and Fukui. Indeed, Herzberg was the most important contributor to the early issues of the *Journal of Molecular Spectroscopy*, a new leader in this field.

Herzberg's life was yet another example of the inadvertent enrichment of North American science through the Nazi policies that expelled Jews. By the time his work on molecular spectroscopy won the Nobel Prize in 1971, he had long since established himself as Canada's leading citizen-scientist — and Germany was out another Nobel Prize.

### Making a Printable Image of the Visualization: The Rise of Computer Graphics in Chemical Physics

After World War II, the possibility of rendering drawings of electron shells and orbitals became more realistic. In England, coming from the chemical side, we find Charles Coulson. He worked with hand-cranked calculating machines and essentially "connected the dots" (various electron point calculations) he had separately plotted. Many of his most influential papers appeared in the British *Transactions of the Faraday Society*. In the U.S., work along these lines from the physics side was pursued by Robert Mulliken, an MIT PhD trained by Slater who spent the bulk of his career at the University of Chicago. Mulliken's work appeared in the *Journal of Chemical Physics*, and indeed this journal, based for most of its life at the University of Chicago, was to feature some of the best work of virtually every scientist mentioned in this chapter who was active after the journal's founding in 1933. In 1966, Mulliken was awarded the Nobel Prize for his work there.

With the increase of computerization in the 1960s, progress was particularly rapid, and, like the spread of computer technology, it was trans-Atlantic. On the Continent, the leader was Per-Olov Lowdin, a physicist at the University of Uppsala in Sweden. Lowdin was something of a human engineer as well, and hit upon the highly novel idea of recycling older, valuable, nominally "retired" scientists. After observing the trend for these scientists to semiretire or overwinter in Florida, he networked together a community of important quantum chemists and physicists. For part of the year, Lowdin works out of the University of Florida, as did Slater, and as have Dirac and Mulliken at Florida State. His ingenuity has resulted in the *International Journal of Quantum Chemistry*, a journal based in both Florida and in Sweden.

An independent, Pittsburgh school of computer-assisted graphical representations arose from the work of Robert Parr, whose work first came to notice while he was at Carnegie-Mellon. Parr was one of the first Americans prominent enough to serve on the editorial board of the leading German journal in this field: *Theoretica Chimica Acta*. John Pople, a British scientist transplanted to Pittsburgh, has in effect sustained Carnegie-Mellon's tradition now that Parr is at Chapel Hill. Pople is the principal author of the "Gaussian" programs. Today, desk-top programs such as Gaussian execute in minutes highly accurate molecular orbital pictures that took months to construct in Coulson's era of adding machines and connecting the dots.

## It Happens in a Flash:
## The Rise of Kinetic and Photochemical Studies

Kinetics is the study of the speed of chemical reactions, and the factors that affect these rates. It is often tied in with the study of the pathways of reactions, a field called mechanistic studies. The commonsense notion that some reaction paths are longer and slower than others is true, but freeze-framing chemical reactions to check on which pathway is being followed is not always easy. A number of developments in photochemistry (the study of changes in chemical in response to light) have proven highly beneficial to kinetic and mechanistic studies. Consequently, for much of the twentieth century photochemistry and kinetics have become intertwined.

### An Explosive Start to the Problem
### of Explaining the Speed of Reactions

The problem of chain reactions was one of the first to be success-fully analyzed in chemical kinetics. By the term "chain reactions" chemists generally mean those combinations or breakdowns that require a little "activation energy," a slight kick start, so to speak, and then proceed very rapidly on their own energy. Two scientists from very different backgrounds and countries came to formulate very similar mathematical and chemical expressions for these chain-reaction phenomena.

The first scientist was Nikolai Semenov, a physicist who literally received his training while the Russian Revolution and subsequent Civil War went on around him. He worked with phosphorous chem-istry, and his materials provide a good example of a chain reaction. Consider the old-fashioned wooden match. Striking the tip just once on a rough surface results in the very rapid inflammation of the whole match head. Continuous scratching of the head against a rough surface, like rubbing two sticks together to start a fire, is entirely unnecessary. In other words, the reaction is not only self-sustaining, it is self-accelerating until the phosphorous of the match head is very rapidly exhausted and the wooden match shaft itself is ignited. While Semenov actually worked with phosphorous vapors, which are even more explosive once a trigger point is reached, the key ideas are the same. In certain reactions, the immediate products of the initial triggering process multiply dramatically, greatly speeding up the reaction.

In later, calmer times, Semenov extended his findings to areas of chemical engineering, particularly to the making of polymers in a chain-reaction fashion. (Recall from the chapter on organic chemis-try the sudden, unexpected, and irreversible changes within the labs of chemists working with clear fluids and hoping for clear crystals when their mixing fluids suddenly became sticky gels or gummy rubbers!)

This work was reinforced by the findings of a former British munitions chemist, Sir Cyril Hinshelwood. He discovered that un-der a variety of fairly ordinary conditions, the combination of hy-drogen and oxygen gases to form water vapor proceeded slowly.

Given the right conditions, however, the result was quite an explosion. (This is exactly what happened to the Hindenburg zeppelin.) Hinshelwood worked under better circumstances than Semenov, and provided a more sophisticated series of chemical and mathematical expressions than the Russian, but along very similar lines. Later in his career, the British scientist extended the findings to certain biological phenomena as well, for example the explosive multiplication of bacteria. In 1956, both the remote Russian pioneer and the elegant British refiner of chemical chain-reaction theory were awarded Nobel Prizes.

## Shining a Light on the Culprit: Demonstrating the Existence of Fugitive Intermediates

A major problem in dealing with chemical transformations is that historically one could be certain of which substances were present at the start of a reaction and which substances were present at the end, but what existed or was generated along the way was much less certain. Differences in speed along the way to an end product had been explained by changes in temperature, pressure, concentration of reactants, and the like. To some degree, each of these factors played a role, but it was long considered that one of the ways each of those physical factors worked was by influencing which intermediate or temporary chemical substances were produced at a given stage of an ongoing reaction. Given chemical intermediates could be isolated from time to time, provided that the scientist could wash out, or cool down, the reaction to isolate them before they could further react. But each attempt at neutralizing the reaction introduced the risk of changing what substances were there inadvertently. Moreover, while not all reactions were quite as explosive as those of Semenov and Hinshelwood, many seemed almost as fast — too fast to stop in time to trap the intermediates.

Occasionally, chemists got lucky and inadvertently created a longer-lived intermediate. These compounds were quite unstable and had substantially different electron charges than expected. Indeed, they were termed "radicals." In the late 1800s and early 1900s, an American organic chemist, Moses Gomberg of the University of Michigan, had more success than most in at least tentatively isolating and identifying these compounds. Gomberg did have some followers such as Francis Owen Rice of Catholic University in Washington during the 1920s and 1930s, but the group of believers in the existence and importance of radicals remained quite small.

It was to take two British scientists and a German outside of organic chemistry to verify and exploit these radical intermediates. The tool of the British scientists, George Porter and Ronald Norrish, was surprisingly simple — a movie-camera-like spectroscope. The underlying principle was also familiar to most people: a bright flash of light triggers a chemical reaction. Lay people regard this as just an element of photography, but Porter and Norrish saw this flash as an opportunity to create radicals and quickly snap a picture of their existence as shown on the spectroscope.

Porter and Norrish would essentially split a strong light beam into two. One beam would cause the onset of the reaction, the other would essentially trip the shutter of the spectroscope, snapping a picture of reaction in a split second. With improvements, Norrish and Porter were able to take pictures of light-stimulated reactions that lasted only a thousandth of a second. Over time, they were able to piece together what amounted to animated spectral films. Evidence of radical intermediates was increasingly abundant, and the intermediate steps along the pathways of many chemical reactions could be resolved. The ideas of Gomberg and Rice concerning the widespread, if short-lived, existence of radicals have been vindicated.

Today, with the use of fairly ordinary light sources, it is possible to observe intermediates that last only a millionth of a second. Studies using conventional light sources have been particularly well financed in biochemistry, owing in part to the photochemical nature of photosynthesis in plants and bacteria, and the role of sunlight in skin cancer and in vision.

Lasers have added tremendously to photochemistry, and have forged alliances between photochemists and optical physicists. Indeed a knowledge of photochemistry has helped build "smarter" or "tuneable" lasers. Lasers can be tuned to flash only certain wavelengths of light, selectively snapping only particularly susceptible bonds. Of even greater interest to physicists, lasers provide a greatly increased intensity of flash. Lasers provide extremely sharp and very fine spectral lines for spectroscopists. (This development won Nobel Prizes for Nicolaas Bloembergen and Arthur Schawlow, a Netherlander and a Canadian, respectively, both later transplanted to the U.S.) Lasers can hyperexcite atoms to become especially reactive or "hot," a phenomenon of interest to an area called plasma physics.

Given these developments, it is not surprising that Norrish and Porter were to win the Nobel Prize in 1967. They shared the prize with a German chemist, Manfred Eigen, who successfully exploited the flip side of excitation: relaxation. Eigen reasoned that the amount of energy it takes to photochemically or electrically excite a compound or create a radical tells us a great deal about its ordinary stable state. Manfred Eigen became an expert on the time it took for an excited compound to return to normal, or to chemically "relax." In this emphasis on a downward shift to a lower energy state, Eigen's work linked photochemistry back to thermodynamics, a longer-established fields within physical chemistry, and made it academically more respectable. Ironically, Eigen's work on "relaxation" kinetics has on its own produced a frenzy of studies.

## A Kind of Dance:
## The Surrounding Liquid as a Chaperone
## Encouraging Awkward Adolescent Molecules to Meet

For decades, solution chemistry and quantum mechanics were not tied in to kinetic studies. This was curious because it was quite clear that, at least in chemistry, the liquid state certainly was conducive to chemical reactions. Moreover quantum chemistry had enabled chemists to learn a good deal about the shape and reactivity of atoms and molecules. More than anyone else, Henry Eyring, a chemist who worked primarily at Princeton and later at the University of Utah (he was a prominent Mormon), tied together these separate themes.

Eyring had a colorful career that involved all three of the regular phases of matter. He originally worked as a mining engineer — the solid phase of his career. He then shifted to doctoral work that involved gas physics and early quantum mechanics. A little later,

Eyring worked with Joel Hildebrand of Berkeley and Farrington Daniels of Wisconsin, both of whom were trained and strongly influenced by van't Hoff, Ostwald and Nernst, and had a strong interest in the liquid state of matter. As Eyring matured scientifically, he too suggested that liquids had a certain structure supported by bonds between the liquid-state molecules. These bonds were weak when compared to most chemical bonds, but were curiously effective and necessary to keep the liquid a liquid. Otherwise, Eyring reasoned, all liquids would instantly evaporate. A Dutch scientist, Johannes van der Waals, was an early European proponent of this view but argued about the way heavy gases and liquid-gas mixtures clung together. Eyring extended van der Waals theory more fully to the liquids themselves.

Eyring reasoned that the ability of one molecule within a liquid to move around and react with a target molecule within the liquid was affected by its ability to break some of the bonds that held the host liquid together. (In other words, the chaperoning liquid had to allow some movement or the boys and girls would never get close enough to dance.) Yet the bonds that held the liquid together had to be strong enough to keep the reacting molecule and the target molecule from escaping as vapors before they had a chance to meet within the liquid and react. (If the chaperones were too ineffective, the boys and girls might separately leave the dance before the music played.) Together with Michael Polyani (a mathematical physicist already famous for his initially disbelieved, but eventually proven, interpretation of the chain-link structure of polymers) and Eugene Wigner (a distinguished quantum mechanic physicist), Eyring proposed certain predictive formulas for the rate of chemical reactions (the success rate of getting the kids to dance). Their joint work on surface potential took into account all three major players in the dance: the boys, the girls, and the chaperone molecules that made up the surrounding solution.

The historical irony of this achievement was that only one of these three collaborators, Eugene Wigner, was to win the Nobel Prize, doing so in 1963. In a pattern common among physicists who involved themselves with chemistry, the prize was not for the discovery most important to chemistry. In an even more startling development, Polyani gave up conventional science and became a sociologist! (He must have been extremely depressed to have fallen that far!) Nonetheless, with Eyring's equations, an at least partly quantum mechanical approach to reaction rates in solutions was on its way. This and further mathematicization by the followers of the British kinetics chemist Moelwyn-Hughes have sustained the study of liquid phase kinetics and the transport of substances within fluids throughout this century.

## Not Quite a Solid, Not Quite a Liquid: The Special Case of Colloids

Physical chemistry was dominated by the three main phases of matter: gases, liquids, and, to a lesser degree, solids. But just as organic chemistry has the awkward problem of polymers, physical chemistry has colloids. Colloid science results from observing an assortment of common, fairly stable, yet not readily explainable mixtures. With colloids, the entrapped or suspended materials do not quite lose their original identity, much as salt does in water. Nor do they quickly fall out of their environment, much as sand stirred into water would. Only relatively long periods of time can separate most colloids.

Colloids can be the mixture of gases in liquids that we call foams. Colloids include the mixtures of thick liquids in thin liquids called emulsions. Colloids can be the mixtures of solids or liquids in air we know as aerosols. All of these, plus gelatins, are definitions of major categories of colloidal suspensions. Shaving cream, mayonnaise, smoke and Jello are everyday examples respectively of each category. Colloid science has also embraced the explanation of phenomena like absorption of fluids by fibers, the creation of froths and bubbles, and the concept of slickness and "wettability." It may surprise the reader that many cosmetics, soaps, and detergents are created by, or work through, colloidal action. Likewise, the cleanup of oil spills or other toxic substances on water requires a good deal of understanding of colloids and surfaces.

Colloid studies have gone in two major directions. The first involves separating out colloids and trying to understand their basic properties. The second exploits colloids that stay in suspension and tries to engineer those suspensions for practical needs.

The basic science approach to colloids in liquids began in the 1860s with Thomas Graham, a Scottish chemist. He observed that unlike most crystalline substances placed in solution, certain starches and gelatins would not pass through a parchment membrane. In terms of today's high school biology, they did not undergo rapid "osmosis." Graham named these substances after the Greek word for glue, doubtless because of their gummy texture. While many nineteenth century scientists would prepare and examine colloidal emulsions, the most significant figures in colloid studies would come from rather afar, and after the turn of the century.

Two men who were to win the Nobel Prize in successive years (1925-26) worked on the problems of the size of colloids, but from very different angles. Richard Zsigmondy was an Austrian who lived, studied, and worked largely in Germany. Although he had a fine education in both physics and chemistry, it was his practical experience in glass factories that led him to work with colloids. He was particularly interested in the properties of colored glasses, and guessed, correctly, that the coloring particles were colloids. Using his connections in the glass trade, he obtained use of the laboratories of Carl Zeiss, a firm still renowned today for superb microscopes. There, by dint of practical tinkering, he invented the ultramicroscope, an instrument with the correct power and lighting angle to actually see the colloids about which other scientists had only been able to conjecture. He went on to study the size and motion of colloids and worked largely with a view towards engineering more useful colloids.

By way of contrast, Theodor Svedberg was a pure scientist at heart, although he was to make many practical advances in colloids. Svedberg's approach to colloids was inspired by the electrochemical tradition of Davy, Faraday, and Kohlrausch and he was among the first to suggest that various electrochemical surface charges held colloids suspended within fluids. Svedberg's approach to colloid study was in the vein of early analytical chemistry: separate the colloids out and then analyze them. Svedberg did this through his invention of the ultracentrifuge. In an ultracentrifuge, a colloidal suspension is placed in a test tube which is then whirled around at tens of thousands to hundreds of thousands of times per minute. Svedberg was able, through centrifugation, to make colloids defy one of their cardinal behaviors and come out of suspension for further analysis.

Two categories of colloidal substances were favored by Svedberg. These were proteins, analyzed at the behest of medical scientists, and polymers, analyzed at the behest of a Swedish government blockaded from natural rubbers during World War I. In both categories of materials, Svedberg discovered a surprising uniformity of size among the colloidal particles. He forced traditional academic chemistry to realize that colloids and polymeric macromolecules (which could also be dispersed as colloids) were areas worthy of serious study and not just confused jumbles too messy to deal with. Svedberg was, in a very real sense, the Staudinger of colloids.

With time, practical engineers in line with Zsigmondy's approach, such as metallurgists, realized that many metal alloys made from molten materials, once thought to be purely crystalline, turned out to have colloidal character as well. Using electrochemically dissolved metals, engineers applied the basic discoveries of Svedberg and Zsigmondy to create various heat-resistant metallic gels for special lubricants. (Most premium motor oils today use these as additives.) Likewise, chemical, pharmaceutical, and food engineers worked with Svedberg and Zsigmondy to develop blenderizers to accomplish just the opposite of what the centrifuge did. Indeed, virtually every biochemist today starts his research by blenderizing some large biological chunks of material to create colloids and then ultracentrifuging out the desired fraction of the colloidal suspensions.

The vapor type of colloid, the aerosol, was popularly discussed in the late 1800s by John Tyndall even before it was able to be readily reproduced in the lab. While Tyndall was well trained on the Continent in mathematics and physics, his greatest claims to fame were as a kind of antisuperstitious, antireligious, antiestablishment radical. He was hired at the Royal Institution by the quietly conservative and deeply religious Faraday, a man almost totally unlike Tyndall save for their mutual interest in the scientific education of the common people. Tyndall systematically observed the beam effects of light passing through smoke, fog, and soot, and made this behavior a kind of test for the colloidal nature of air pollution. He used this knowledge in his studies of mine explosions, which he showed were due not only to methane gas (Faraday's finding initially) but to suspensions of coal dust in the air as well.

Practical attempts to deliberately create aerosols focused for much of the nineteenth century on atomizers, such as those used for perfume sprays. By the early twentieth century, pressurized gas was being used as a propellant, and the modern spray can came into use in the 1950s. Ironically enough, the search for better, environmentally safer spray cans has gone back to the atomizer, given the damaging effects of some of the propellants. Tyndall would have approved.

More than any other national group of scientists, Americans have tied together the themes of surface behavior and colloids. Largely through the work of Irving Langmuir and his colleagues in industrial research, Katherine Blodgett and Raymond Fuoss, a major focus has been on the thin layer or interface between the colloid and the surrounding solvent. Langmuir was a Columbia University graduate who took his graduate degrees in Germany under Nernst. Nernst's electrochemical outlook and propensity to make money from research appealed to Langmuir. Within a short time after his return to America, he began a long and very fruitful career with General Electric, winning the Nobel Prize in 1932. A curious example of his wide expertise in colloid behavior was his improvement of the smokescreens the U.S. Navy deployed during World War II. Today, the American Chemical Society's journal for colloids and surfaces is named after him.

Columbia University was also the base for other American pioneers, such as Victor LaMer, who tied in many aspects of quantum theory and solution chemistry to colloid behavior. LaMer's most influential colloid protégé at Columbia was Milton Kerker, who became a leading aerosols expert at Clarkson.

The Columbia-Clarkson group was to link with another power in colloids: Lehigh University. Albert Zettlemoyer graduated from Lehigh University, took his PhD at MIT, and returned to Lehigh after a stint in industry. His most important contributions include advances in a colloidal suspension on which this book is particularly dependent: the improvement of printer's ink. Lehigh was particularly famous for a series of special institutes run by Zettlemoyer that gave some identity to the diverse body of special interests that constitute the field today. LaMer and Kerker participated in these institutes and helped shape one of the most influential journals in the field: the *Journal of Colloid and Interface Science*.

Today's studies of colloids tend to be mildly split along traditional "wet" vs. "thin layer" lines. Traditional lines include colloids that have a polymeric base (e.g., latex paints), a biochemical aspect (e.g., proteins in solution), or other conventional application (e.g., detergents). The thin layer or interface approach views the surface composition and charge or other physics aspects of the colloidal suspension or film as most important. This has come to include items as diverse as biological membranes and computer chips.

## SORTING OUT THE JOURNALS

### Journals of General Physical Chemistry

While the distinctions are rapidly fading, journals of physical chemistry — as opposed to chemical physics — tend to have more papers from scientists trained as chemists, tend to have more liquid-phase papers, and are more molecular than atomic in any spectroscopic studies they feature. While physical chemists are a rapidly increasing component of national chemical societies, fewer of those societies sponsor separate journals for physical chemistry than is seen for organic chemistry. Moreover, there is generally less space in the major national general chemistry journals devoted to physical chemistry than to organic. Nonetheless, three of the leading physical chemistry journals are society sponsored, two by general chemical societies, one by a special society for physical chemistry. (See Figures 1 and 2.)

The international leader is the *Journal of Physical Chemistry* from the American Chemical Society. It leads in share of American papers, impact factor, total number of papers, and is arguably the most reasonably priced. It is nonetheless highly international, with almost 40% of its papers coming from abroad.

The second choice is less clear, given the strong traditions of both the British and German efforts in physical chemistry. There are arguments for either option. The British option, the *Journal of*

FIGURE 1

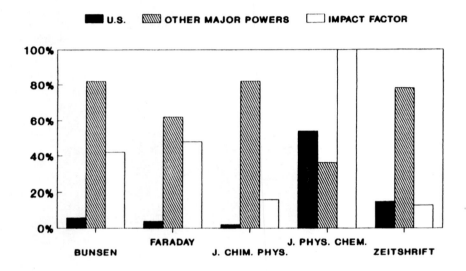

## Contributors and Relative Impact
## Leading Physical Chemistry Journals

■ U.S.    ▧ OTHER MAJOR POWERS    ▢ IMPACT FACTOR

*the Chemical Society, Faraday Transactions* is, however, less complicated. All of its papers are in English, their number is higher, and its impact factors are marginally better. Most importantly, this one title from the Royal Society of Chemistry gets virtually all of the best papers from the U.K. and the Commonwealth, particularly since it ceased separate "I" vs. "II" editions for the chemical vs. physical emphasis.

The total German effort is formidable but is unfortunately split three ways, with only two titles internationally viable. The *Berichte der Bunsengesellschaft fuer Physikalische Chemie*, from VCH for the Bunsen Society, clearly invokes one historical figure, even as Ostwald and van't Hoff are claimed by two versions of the *Zeitschrift fuer Physikalische Chemie*. The West German version from Oldenbourg in Frankfurt is the title that is analyzed in the figures, not its formerly East German rival from Akademische Verlagesell-

FIGURE 2

## Comparison of Number of Papers, Costs, and U.S. Market Penetration

■ NUMBER OF PAPERS ▨ COSTS
□ LIBRARY SUBSCRIBERS

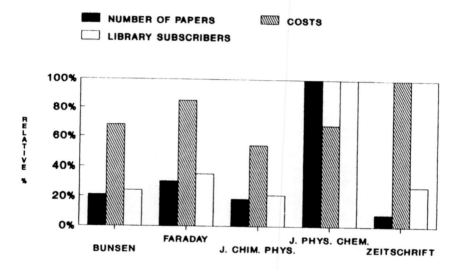

schaft in Leipzig. Moreover, the western version is much more likely to survive German reunification in the long run. There is some residual use of German in all three of these journals, and their comparative costs are not quite as favorable as the British entry. Nonetheless, they do have some advantages. In particular, the two major German journals attract a greater total share of their papers from other scientifically advanced countries (many of the Commonwealth papers come from economically disadvantaged countries) and virtually all of the non-German national papers in the *Zeitschrift* and *Bunsen* are in English. If a librarian is forced to choose only one title from among the German offerings, the figures show *Bunsen* performing better in most measures.

The remaining title in this section is French, and still features some French language papers. Nonetheless, the *Journal de Chimie et de Physicochemie Biologique* has merit in its explicit ties to biochemistry, and is a good choice in comprehensive collections, or those with a life sciences accent. In the latter situation, *Biophysical Chemistry* from Elsevier, and *Biophysical Journal* from Rockefeller University Press, are also strongly recommended.

General physical chemistry is nicely served by the *Annual Review of Physical Chemistry* for traditional reviews, and by the *Faraday Discussions of the Chemical Society*, a symposium journal from the Royal Society of Chemistry. Physical chemistry with a biological slant is covered by the *Annual Review of Biophysics and Biophysical Chemistry*. The *Annual Reviews* are published by the not-for-profit Annual Reviews, Inc.

### Journals of General Chemical Physics

Journals of chemical physics have more papers from physicists than from chemists. There is a greater prevalence of theoretical or quantum chemistry, and experimental work involves more gas and solid phase papers. Highly detailed spectral analysis at both atomic and molecular levels is common, particularly when the electron pathways are stressed as in strong magnetic fields or in laser excited states.

With the consolidation of the *Journal of the Chemical Society, Faraday Transactions* into a single journal with a preponderance of chemically-oriented papers, the field of chemical physics is now populated largely by for-profit titles with a single major society entry. (See Figures 3 and 4.) However, that single entry, the *Journal of Chemical Physics* from the American Institute of Physics, is a powerful competitor with serious advantages in virtually every measure. These are likely to remain in light of the journal's strong tradition and its continuing editorial ties to such research powerhouses as the University of Chicago.

Its closest competitor offers a single, but significant advantage. *Chemical Physics Letters* from Elsevier has become the speed-of-publication leader in the field, and appeals to both chemists familiar with *Tetrahedron Letters* and physicists who use *Physical Review Letters*. Indeed, this brief-paper, rapid communications child of the conventional journal entitled *Chemical Physics*, has eclipsed its

FIGURE 3

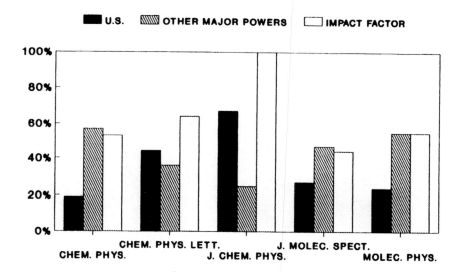

## Contributors and Relative Impact
## Leading Chemical Physics Journals

■ U.S.    ▨ OTHER MAJOR POWERS    ☐ IMPACT FACTOR

parent in prominence, and is the best second choice, despite its
steep cost.

The remaining journals are competitive on their own intermediate
level. *Molecular Physics* from Taylor and Francis is somewhat
more reasonable financially than *Chemical Physics*, as is the *Journal of Molecular Spectroscopy* from Academic. Both of these titles
are somewhat narrower in scope than *Chemical Physics*, but the
specialties they do emphasize are quite essential to virtually all
chemical physics collections. *The Journal of Molecular Spectroscopy* does have an advantage of doing double-duty in schools where
analytical chemistry is offered, since some molecular spectroscopy
papers are valuable for that specialty. Most large collections should
strive for all three if possible.

While most physics oriented collections would already have
*Physical Review A* from the American Institute of Physics, chemis-

FIGURE 4

# Comparison of Number of Papers, Costs, and U.S. Market Penetration

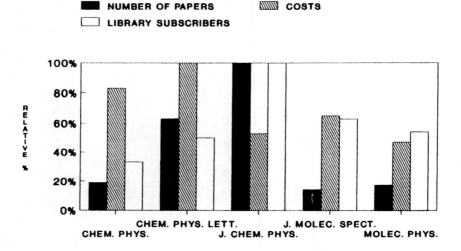

try collections should make a point to take their own duplicate. The journal appears semi-monthly. On the first of each month, papers in atomic, molecular, and optical physics are featured. On the fifteenth of each month, papers in statistical physics (an area with close ties to thermodynamics) and fluids appear.

The leading review in this area is from Wiley: *Advances in Chemical Physics*, an irregular hardbound series.

## Journals of Thermodynamics

The wide assortment of thermodynamics titles available today is due to the diverse nature of disciplines with an interest in the field. The notion of a clear-cut leader among all thermodynamics journals may not be appropriate. The best choice for a given collection may well depend on either a journal doing double duty with another specialty in the institution, or otherwise closely matching an institutional emphasis. (See Figures 5 and 6.)

Conventional chemical engineers, particularly those working with hot liquids and molten materials, will need to consult the extensive studies in journals such as *Fluid Phase Equilibria* from Elsevier. While the cost of a subscription would be steep, the materials discussed are generally economically important and processed in large quantities. The clientele for journals such as *Fluid Phase Equilibria* tend to think on a different scale of financial terms than the librarian in cost/worth equations. It is the best choice in a department of chemical engineering. A related title, not pictured here, is the *Journal of Chemical and Engineering Data*. While it is less expensive, coming from the American Chemical Society, it is also less thermodynamically focused.

In recent years, the thermal analysis of solids has become in-

FIGURE 5

## Contributors and Relative Impact
## Leading Thermodynamics Journals

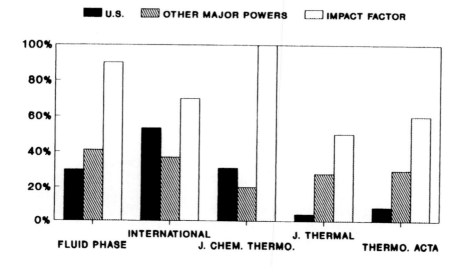

FIGURE 6

## Comparison of Number of Papers, Costs, and U.S. Market Penetration

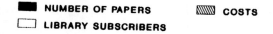

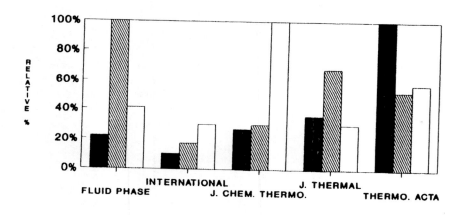

creasingly important, and the *Journal of Thermal Analysis* from Wiley seems to have a lead, although not an exclusive focus, in papers of this type. It is the best choice in programs that have a materials science department. Nonetheless, many papers that use thermodynamic properties analytically are also to be found in Elsevier's *Thermochimica Acta*. *Thermochimica Acta* places Elsevier in an unusual situation this time: it has the most papers in the field (a characteristic of Elsevier) without having the highest price (usually a concomitant Elsevier characteristic). It is probably a better choice for programs that also have a strong analytical chemistry component.

In small schools, particularly those in the basic sciences as opposed to engineering specialty programs, two titles are clear first choices. Straightforward programs in physical chemistry within chemistry departments will probably prefer the *Journal of Chemical*

*Thermodynamics* from Academic Press, while straightforward physics departments will lean towards Plenum's *International Journal of Thermophysics*.

Very large collections, particularly in engineering schools where thermodynamics is often emphasized, would take all of the titles mentioned above. They may also wish to consider the *Journal of Non-Equilibrium Thermodynamics* from Walter de Gruyter, and *CALPHAD: Computer Coupling of Phase Diagrams and Thermochemistry* from Pergamon.

## Journals of Solution Chemistry and the Fundamental Electrochemistry of Solutions

Selection in this area is fairly clearcut. The *Journal of Solution Chemistry* from Plenum is the leader. Indeed, this journal has concentrated many of the more fundamental solution chemistry papers that used to appear in electrochemistry, thermodynamics, and kinetics journals. A good second choice is Elsevier's *Journal of Molecular Liquids*. It has a strong emphasis on denser and more complex fluids, like large organic solvents and polymers.

The choice of electrochemical journals is also obvious, once it has been decided that the small amount of solution chemistry remaining in these journals is necessary for a particular institution. Given its impact and low costs, the *Journal of the Electrochemical Society* is the first choice, followed by *Electrochimica Acta* from Pergamon. Any argument for inclusion of these latter two journals is strengthened if there is a materials or surface science working group among the clientele, since electrochemistry is now used heavily to create specially tailored coatings and thin solids. If an analytical chemistry emphasis is present, the *Journal of Electroanalytical Chemistry and Interfacial Science* from Elsevier or the new VCH title *Electroanalysis* is recommended instead, since many solution studies involve electrode analysis.

## Journals of Quantum Chemistry and Molecular Modeling

The current popularity of manuscripts in the intertwined areas of quantum chemistry, molecular modeling, and chemical computer graphics is very high. There is scarcely a month in which new software or new computer speed and capacity have not made a new calculation or visualization possible. Important papers will be found in the major general chemistry journals, particularly the *Journal of the American Chemical Society*, and major chemical physics titles, particularly the *Journal of Chemical Physics*. Nevertheless, three journals dealing primarily with the theoretical analysis of chemical substances have become well established. These are supported by two journals dealing with the software tools used to execute these studies.

There is no clearcut leader among the three major conceptual journals. (See Figures 7 and 8.) All three are of high quality, with some small advantages or nuances that make differentiation possible when it is not feasible to take them all. The *International Journal of Quantum Chemistry* from Wiley has the most papers, including many from U.S. authors, and is the most reasonable financially

FIGURE 7

## Contributors and Relative Impact
## Leading Quantum Chemistry Journals

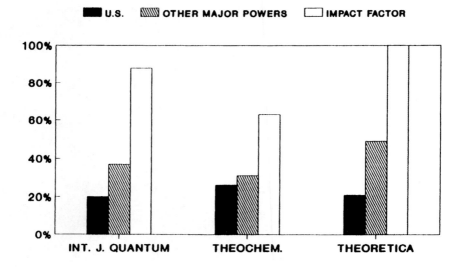

FIGURE 8

# Comparison of Number of Papers, Costs, and U.S. Market Penetration

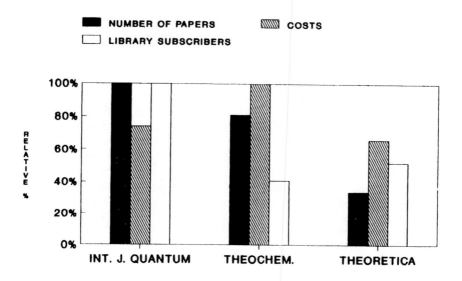

given the numbers of those papers. *Theoretica Chimica Acta* has the highest impact factor, but its cost-per-paper ratio is severe. By way of partial explanation, it should be noted that *Theoretica* publishes the longest papers in the field as well.

Yet *Theochem*, the third conceptual journal, is not without an advantage. It takes more papers in organically-oriented theoretical chemistry. Libraries that take the *Journal of the Chemical Society, Perkin Transactions II* or the *Journal of Physical Organic Chemistry* might give this journal some preference. It should be noted that *Theochem* is actually a subsection of a classic Elsevier omnibus title that the strongest collections must have: the *Journal of Molecular Structure*. The parent title includes many spectroscopic investigations which reinforce theoretical studies nicely. Likewise, serious collections in theoretical chemistry should also be sure to take the *Journal of Molecular Spectroscopy* discussed in the general chemical physics section of this chapter and within the chapter on analytical chemistry.

Two journals (not depicted in the figures) complement the three major theoretical titles closely: the *Journal of Computational Chemistry* from Wiley, and the *Journal of Molecular Graphics* from Butterworth. Here the leadership is clear. The title from Wiley is much better established, very reasonably priced, and includes both software and substantive studies. Yet the *Journal of Molecular Graphics* has the advantage in somewhat more stress on depicting the large molecules (proteins, nucleic acids, etc.) favored by biochemists and molecular biologists.

Librarians and data processors within the indexing-abstracting, on-line, and CD-ROM industries may also wish to consider the *Journal of Chemical Information and Computer Sciences* from the American Chemical Society. While the graphic portrayal of chemical structures is not its primary purpose, it has information along these lines as well as the best articles on the encoding and retrieval of properties information in chemistry overall.

### Journals of Chemical Kinetics

Selection in this field is very straightforward: the *International Journal of Chemical Kinetics* from Wiley is the leader for initial research reports, and *Progress in Reaction Kinetics* from Pergamon is an appropriate source of reviews.

### Journals of Photochemistry and Laser Chemistry

Selection in this field has become complicated. A three-way divergence in research emphasis has appeared, and competition is fierce. (See Figures 9 and 10.)

Over the last several years, the traditional leading journal *Photochemistry and Photobiology* (the organ of the American Society for Photobiology, from Pergamon) has become very heavily committed to the biology end of the field. It remains absolutely the first choice for that audience, but is no longer the first choice for the physical chemistry or chemical physics segment of photochemistry. That title has clearly devolved on the *Journal of Photochemistry and Photobiology A - Photochemistry* from former rival publisher Elsevier.

FIGURE 9

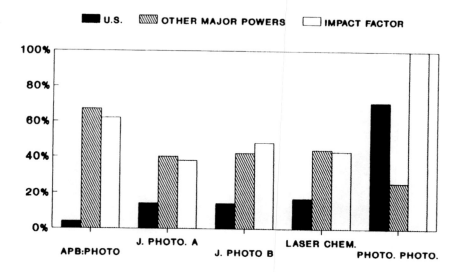

## Contributors and Relative Impact
## Leading Laser & Photochemistry Journals

■ U.S.   ░ OTHER MAJOR POWERS   ☐ IMPACT FACTOR

*Photochemistry and Photobiology* has recognized this very recently and has announced a new section exclusively for mainstream photochemistry, with sponsorship from a coalition of photochemistry societies. Is it too late to recapture the photochemistry mainstream? Pricing may play a role in the long run. At this time there is a significant difference in cost-per-paper favoring of Pergamon. But will chemists pay much of anything for a journal so biological?

Nor is this the only Elsevier raid on Pergamon's flagship in the lucrative biophysical market. There is also the *Journal of Photochemistry and Photobiology B - Biology* from Elsevier. While it has captured the endorsement of the European Society for Photochemistry, it still has substantially fewer papers, lower impact, and higher per-paper prices than *Photochemistry and Photobiology*. It remains a second choice for biophysically oriented collections, and can be omitted in most chemistry collections.

FIGURE 10

## Comparison of Number of Papers, Costs, and U.S. Market Penetration

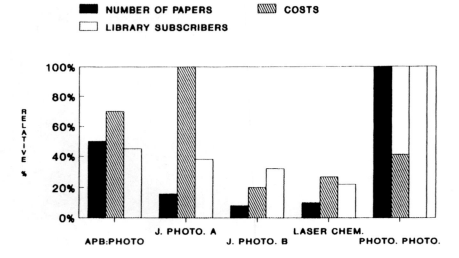

There remains yet another split between journals which emphasize more traditional photochemistry and those which are more heavily committed to laser chemistry. Two appropriately named journals have focused on the laser subspecialty market: *Laser Chemistry* from Harwood, and *Applied Physics B - Photophysics and Laser Chemistry* from the German publisher Springer. Both compete for manuscripts with the relatively new sections within the *Journal of Chemical Physics* and the *Journal of Physical Chemistry* now set aside for this hot field. At this time, both probably attract more attention from the physicists' side of the physical chemistry community, although *Laser Chemistry* might be expected to migrate closer to the chemists over time. The first title clearly belongs in many physics collections, the second very likely so. Physical chemists should take *Laser Chemistry* preferentially.

The leading review journal for the field is the *Advances in Photochemistry*, a hardbound series from Wiley.

## Journals of Colloids

There are many journals taking colloid science papers today. There is, however, a hierarchy among the journals and certain collateral subject interests within some titles that will aid selection. (See Figures 11 and 12.)

In the last few years, *Langmuir*, from the American Chemical Society, has emerged as the colloid leader. It presents us with the usual ACS advantages of a high proportion of highly cited U.S. papers at a reasonable price. It also is a good title for the surface science-oriented physics collection and has a more sophisticated, academic angle on colloids than most of the other titles in this collection.

The *Journal of Colloid and Interface Science* from Wiley, however, has lost little of its attraction for working colloid engineers. This long-established title has the most papers, the second most

FIGURE 11

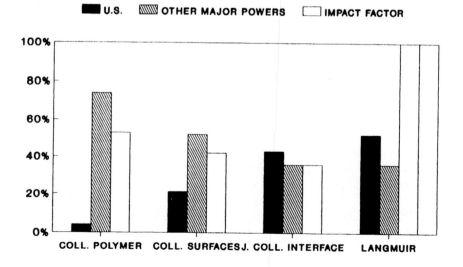

## Contributors and Relative Impact Leading Colloid Chemistry Journals

FIGURE 12

## Comparison of Number of Papers, Costs, and U.S. Market Penetration

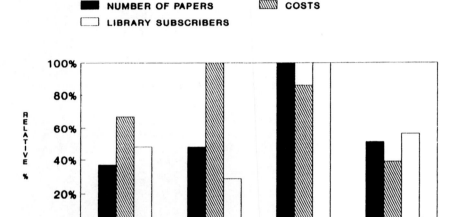

U.S. authors, and second-best impact factor. It has among the "wettest" mix of colloid papers (emulsions of solids or liquids in other liquids) and yet has not lost touch with modern surface science developments. This is a second place title that is almost as necessary as the first place finisher, particularly in the chemical engineering collection.

The remaining two titles are also of high quality, but offer a good subject differentiation on which to make a choice. *Colloid and Polymer Science* from Steinkopff joins two fields (and two separate formerly German-language journals) that were both historically regarded as scientific bastards. That historical linkage is not entirely irrelevant today, since many polymers are important for surface science or in emulsions. The mix of colloid-to-polymer papers is close to 50% each. The subscription cost is somewhat more tolerable for *Colloid and Polymer Science* than for *Colloids and Surfaces*. *Colloids and Surfaces* from Elsevier, however, is more closely comparable in its subject mix to *Langmuir*, the field's current leader. It also has a higher proportion of U.S. authors. If both journals cannot be taken, *Colloids and Surfaces* makes better sense in collections that serve a physics community, *Colloids and Polymers* for collections that serve chemical engineers.

The leading review in this area is *Advances in Colloid and Interface Science* from Elsevier.

# Chapter 5

# Journals for Physics at the Very Small and Very Large Scales of Matter

## BACKGROUND

Scientists have long had an urge to isolate and identify the smallest units of matter and to understand how those units relate to one another. In almost every generation since the turn of the last century, there have been advances that forced a revision of the organizational chart of matter. Successive revisions are usually forced by the finding of smaller and smaller fundamental units. Yet the larger units charted earlier continue to be of scientific interest. As we have seen from the last chapter, molecular physics did not die because of findings about the atom, nor would atomic physics die, as we discovered subatomic particles. In fact, as we continue to find finer levels of matter, each level gives rise to specialized journals. But the old "new journals" neither die, nor fade away. There is a good deal of basic science remaining at the larger atomic scale, and engineering applications at the larger scale are already on the way. A final irony of the hunt for ever smaller particles is that the study of the microscale of matter often illuminates the study of the gigantic bodies and energetic outpourings seen in deep space, and that a consideration of the cosmos tells us something about the inner workings of atoms. It is not unusual for astrophysicists to want journals of particle physics, and vice versa.

### The Two-Fold Way:
### Experiments vs. Highly Mathematical Theory

The means of exploring the world of the small is partly through equipment and experiments, and partly through highly mathematical and theoretical conceptualizations. Contradictions between the two sometimes resulted in victories for one approach or the other. But in most cases, the atom or particle under discussion surprised and confounded both approaches for some time. Often, the attempt to confirm the theory by experiment gave results that could only be interpreted by more theory. Symbiosis and synergy are the best descriptions of their relationship, and should be apparent in both the history of the field that follows and in the mix of topics favored by many of the journals discussed.

### *Because of Thomson and Rutherford, the Atom Becomes More than a Convenient Figure of Speech*

Atoms were talked about long before they were understood. Indeed, although chemists like Dalton are the fathers of the modern chemical theory of the atom, many chemists well into this century did not feel that a close look at atoms was important to do chemistry. While this would change somewhat with the arrival of quantum chemistry, physics had pretty much claimed the serious study of atoms by then.

Radioactivity was the principle inducement for atomic experiments and it captivated the academic establishments in France, Germany, and England during the early 1900s. In France, Pierre and Marie Curie, and Henri Becquerel were to win the 1903 Nobel Prize in physics for their discoveries of the ray-spewing nature of elements like radium. Some rays emanating from the elements seemed to be particularly energetic, and were as penetrating as the high-voltage X-rays discovered in Germany by Roentgen (the first Nobel Prize in Physics, 1901). Others seemed not so strong, and could be stopped more readily or even bent by electrical fields. From the start, there was a controversy. Were these rays or beams composed of tiny bits of matter, typically called particles, or a type of electromagnetic wave energy, like light? This is a conceptual dualism that is difficult to handle even today. Evidence that some rays could be primarily energy came from all sorts of scientific phenomena: electricity, magnetism, spectroscopy, and the practical experiences of chemists who felt most comfortable with an atomic model that had energy connected with it to explain the heat and light given off by chemical reactions and radioactive substances.

J. J. Thomson, a Cambridge University physicist at the turn of the century, was one of the first to propose a model for the atom that helped explain radioactive beams. In Thomson's model, the electrons were all-important. The electrons had four special properties. First, they were negatively charged, a fact that was clearly provable by their attraction for opposite, magnetically positive targets. Second, since everyone agreed that electrons were very light, and that each element had only a specific number of electrons, there had to be many of these lightweight atoms packed very tightly together in order to account for the weight of matter. Third, the electrons coated the surface of the remainder of the atom which Thomson took to be a hollow sphere. Fourth, electrons remained at rest until they popped off their spheres to flow in an electric current, to participate in a chemical reaction, or to become part of a radioactive beam.

Thomson was partly right. We now know that Thomson's type of radioactive beam is in fact composed of beta rays or beta particles — electrons behaving in a certain way that happens only in radioactivity. Since Thomson's model certainly agreed with much electricity and chemistry, he went on to win the 1906 Nobel Prize. By way of confirmation of some of Thomson's claims, the amount of the negative charge on the electron was soon demonstrated by an American, Robert Millikan of the University of Chicago, who was to eventually win the Nobel Prize in 1923. But a big surprise to those who believed unreservedly in Thomson came from a most unexpected source: a farmer's son from New Zealand, Ernest Rutherford.

Rutherford was a foreign exchange student and then eventually a graduate assistant to Thomson at Cambridge University. Like Thomson, he was fascinated by radioactivity but unlike him, he made no particular fetish of the electron. After his doctoral training he took posts in Canada and then back in England, and both countries, along with New Zealand, can claim him as a scientific hero. One of his early findings was that not all the rays involved in radioactivity were negatively charged, as Thomson's electron rays clearly were. Indeed with the help of the negatively charged ends of magnets, he was able to demonstrate that "alpha" rays or particles were positively charged.

Handy with equipment, Rutherford and a German-born graduate student named Geiger later set up experiments that included a variety of radiation detectors. In a darkened room, the detectors were designed to spark when radiation flashed on them. Rutherford had set up a source of his new positive alpha rays or alpha particles. His target for the rays was a very thin metal foil. Since opposite electrical charges attract, Rutherford and Geiger wanted to see if the positive particles in his alpha beams were attracted to the negatively charged electrons that were supposed to be resting on the surface of the atoms in the foil. If the positive beams were attracted, they would be stopped, and there would be no sparking of the detector behind the target. Rutherford and Geiger fully expected a sparkless day.

Surprisingly, the overwhelming majority of the beams shot right through the target and sparked the detector behind it. Rutherford and Geiger asked themselves, how could these beams virtually all be missing the tightly packed atoms in the foil? Every one of the few remaining beams sparked detectors around the room at odd angles. Why so? Indeed, some sparks came blasting backwards rather dramatically. There were no sparkless events at all! Clearly, no positive beams had come quietly to rest on the foil. The few ricochets were particularly energetic and frightening. Rutherford realized that during these few but unmistakable ricochets, his beam was hitting something solid—a hard "nucleus" so to speak.

These few but fierce collisions between positive beams and foil involved quite a repulsive force, given the energy of the ricochets. Rutherford reasoned that the strong repulsion of the positively charged beam must be due to a strongly positive charge on the nucleus, since matching electronic charges are strongly repellant.

Rutherford further proposed a novel explanation for the much larger number of particles that shot right through the target. While the nucleus is very hard, he reasoned, it is also very small. Most of the beams that were aimed to hit the foil actually traveled through empty space within the foil. The small ratio of "hits" to "misses" were indicators of just how small that nucleus was compared to the empty space of the atom. Most of the emptiness of the atom, according to Rutherford, was not inside Thomson's supposed hollow spheres, it was outside the hard little nucleus. In direct contrast to Thomson's idea of closely packed atoms to account for the weight of matter, Rutherford's atoms were surprisingly far apart from nucleus-to-nucleus, with virtually all of their weight in the highly compact nucleus. With time, the positively charged particles that constituted the nucleus came to be known as "protons." (The neutrons that were also in the alpha particles Rutherford used would not be discovered for another thirty years.) With the confirmation of the reality of the electron and the proton, the trend of seeking to identify atomic particles and characterize their behavior now became firmly entrenched as a goal of advanced physics.

Rutherford went on to win the 1908 Nobel Prize. Geiger went on to develop one of the few pieces of nuclear instrumentation familiar to the general public: the Geiger counter. A question remained: if the electrons weren't in the nucleus, where were they?

## Max Planck Reluctantly Introduces Quantum Radiation: A "Hopscotch" Approach to Energy Emission

Max Planck was a student of the great spectroscopist Kirchhoff mentioned in the chapter on physical chemistry and chemical physics. Planck achieved a good deal of renown in his own right as the leading German university researcher in thermodynamics. For years he had tried to explain why there were gaps between the lines of Kirchhoff's spectra of compounds. Why shouldn't there be a continuous spectrum, constituting a solid band? He also sought to explain why the energy he saw emitted in his own thermodynamics work seemed to come out only at certain wavelengths. Why, for example, was the conventional experience of stove heat largely of infrared wavelengths and not a complete mixture of all the wavelengths?

In a celebrated address, he admitted defeat. Planck explained that despite the fact that it intensely offended his sense of order, it was clear that energy was emitted only in certain quantities ("quantums" or more properly "quanta") and only at certain widely spaced wavelengths. These units of energy or their multiples always matched certain units or multiples of spacings between the spectral lines. It was also clear to him that in energized situations there was an abrupt jumping of these units of energy over the gaps from one spectral line to its next multiple line and not any gradual transitions between the lines. Indeed, Planck provided a factor to calculate these multiples of energy and spectral gaps that is now known as "Planck's constant."

### Bohr Links Planck's Quanta with Electron Pathways

It was up to a young Danish scientist, Niels Bohr, to put Planck's notions of the spacings of these quantum jumps together with the spacing of the trackways of the electron. After the completion of his PhD, Bohr went to England on a special fellowship designed to expose Danish scientists to the scientifically more advanced British physics community. The grant was initially for one year, and Bohr managed to be accepted as a post-doctoral student by the already famous J. J. Thomson. This collaboration went badly. Bohr did not find Thomson's idea of resting electrons very satisfactory. The electrons he knew just seemed too fast and too busy to be resting most of the time. Having heard this criticism out, Thomson basically invited Bohr to leave, but this left Bohr with unexpired time on his grant.

Bohr moved on to Rutherford's lab in Manchester and at once struck up an amiable relationship. Most pertinently, it occurred to Bohr that it made just as much sense to have the electron always moving as having it mostly at rest. But where was the electron going that it never actually left the atom altogether? Was the electron running in circles, so to speak? Bohr realized that the electron was orbiting the nucleus and the orbits were those narrow but widely spaced pathways that corresponded to Kirchhoff's spectral lines and Planck's quantum multiples. This idea was duly published in *Nature*, the prestigious British journal of science, and was widely accepted. In a very short time, Bohr became a national hero in Denmark and, indeed, won an honor that had tremendous implications for the next generation of atomic physics.

### A Brewer's Palace Becomes a Center for Quantum Theory

It seemed that the head of the prosperous Carlsberg brewery in Copenhagen had, upon his death, willed not only a fabulous amount of cash, but also his own palatial home to be freely used by "Denmark's most distinguished scientist" until his or her retirement. Despite his robust youthfulness—Bohr was built very much like a large American football player and his brother was, in fact, on the Danish national soccer team—Bohr was judged as best fitting this description. While the mansion had no facilities for experimentation, it provided ample and attractive office space and seminar rooms. Moreover Bohr had a budget that allowed him to recruit and subsidize young scientists and hold seminars on a scale that surpassed most European universities. Bohr not only had the resources but an energetic and sympathetic disposition that quickly identified and encouraged young talent. (Indeed, in a manner totally foreign to the rigid German university social structure, young Bohr socialized on a daily basis with his equally young followers, and on one occasion was stopped at night by the Copenhagen police as he scaled the walls of a bank on a dare!) No less than eight future Nobel Prize winners were guests of Bohr's institute at one time or another, and not only did Bohr himself win the prize in 1922, but his son, Aage Bohr, would win it in 1975.

Bohr's institute had tremendous strength in theory and yet maintained a remarkable willingness to revise its thinking without any sense of embarrassment when hearing of experiments that contradicted it. His group did not shy away from the many inconsistencies that came up in the early years of quantum physics. Not too surprisingly, the most common inconsistency came from trying to decide between a particle vs. wave nature for the electron.

## Pauli Says the Electron Behaves Like a Real Particle

One of the first challenges to Bohr's model came from the experimental findings of workers like Philipp Lenard and Johannes Stark, two Nobel laureates working at German universities and institutes. Their work with X-rays and strong magnetic fields showed that the proposed electron pathways within which, and only within which, electrons were supposed to orbit could be clearly, if very finely, split into parallel tracks. In which tracks were the electrons? they asked.

One of Bohr's Swiss students, Wolfgang Pauli, resolved this dilemma. Pauli worked out rules which suggested that no two electrons could occupy exactly the same track, unless they also had differing spins. Indeed, the notion of a particle that both revolved around the nucleus while spinning like a top introduced the notion of "angular momentum" to atomic physics and was fruitful for studies of other particles as well. Pauli suggested that strong external fields, such as those applied by Stark and Lenard, basically separated electrons that were travelling the same path into some electrons spinning one way, and other electrons spinning in the opposite direction. Pauli's work strengthened the case for the electron as a real particle and not just a unit of energy, because it was hard to imagine an entity without substance spinning around. (Pauli was eventually to win the 1945 Nobel Prize for his wide-ranging particle work.) Yet the idea of wave energy or "field effects" operating on the atom was not to be discounted entirely.

### But a Trio of Bohr's Other Students Shows that the Electron Also Behaves as a Wave

Pauli's explanation proved satisfactory until a French physicist, indeed, a wealthy nobleman, Count Louis de Broglie, published his doctoral dissertation. In it, de Broglie suggested that electron pathways were spaced in a way that suggested something like musical chords. He called the pathways "pilot waves" and demonstrated that the quantum jumps suggested by Planck also fit the behavior expected of the crests of successive electromagnetic waves. While de Broglie's theory was not mathematically rigorous, his observations were clearly way too much of a coincidence to be dismissed lightly. Indeed, de Broglie won the 1929 Nobel Prize for his work. De Broglie's challenge represented yet another intrusion of the developing notion of wave theory on atomic particle physics. By way of response, Schrödinger, an Austrian, and Heisenberg, a German, used mathematical techniques very different both from de Broglie's and from each other's. They separately published papers in rival physics journals, with Heisenberg favoring the *Zeitschrift fuer Physik* and Schrödinger appearing in the *Annalen der Physik*. These papers, which appeared in the same week, accommodated much of de Broglie's view without rendering the electron an entirely disembodied packet of energy or wave front. Indeed, Heisenberg was to argue that electrons were not susceptible to being pinpointed at any given instant, although it was clear that they ran exclusively astride certain energy levels that corresponded to de Broglie's wave crests. These wave crests in turn corresponded to the spectral lines of Kirchhoff, the quantum energy levels of Planck, and the electron orbits of Bohr — at least for the simplest atoms. For a time it was hoped that a certain reconciliation of the findings of spectroscopy, particle physics, and electromagnetic wave theory might be accomplished in a single model. X-ray probes of the inner electron shells of more complicated atoms by Karl Siegbahn of Sweden (Nobel Prize, 1924), bolstered this developing consensus. (Indeed, Siegbahn's son Kai won the 1981 Nobel Prize for developing electron spectroscopy even further.) But this total resolution of views was not to be, despite the fact that Heisenberg was to win the 1932 Nobel Prize and Schrödinger the 1933 Nobel Prize, both for their wave findings.

## Einstein Introduces Field Theory to Bohr's School

This time the model was disturbed by the findings of a young German working in Switzerland named Albert Einstein. For years his general and special theories of relativity had met with skepticism. But while Einstein's general theory of the equivalence or transferability of matter into energy was less understood, his special theory concerning the speed of light and the deflection of light under a strong gravitational field had already achieved dramatic astrophysical confirmations.

Moreover, Einstein had made an important early discovery concerning the absorption of light by metals, which then gave off a flow of electrons, a phenomenon now termed the photoelectric effect. (Indeed, Einstein's 1921 Nobel Prize was for the discovery of this effect, not for his more controversial relativity theories as is commonly supposed.) Einstein found that the energy transfer involved in this early "energy field" experiment involved just the sort of quanta that would be predicted by Planck's thinking. Indeed, the packets of light energy that Einstein discussed as colliding with the electrons came to be called "photons," and in later years, it would come to be realized that these units of energy also qualified as real particles, another puzzling example of wave and particle duality.

It was felt that sooner or later, given Einstein's continuing successes, his remaining relativistic views would have to be taken into consideration in looking at the Bohr atom. The British physicist Paul Adrian Maurice Dirac, yet another of Bohr's visitors, showed that it was possible to reconcile Einstein's approach as far as electrons were concerned, but that the calculations would be even more exact if the existence of hidden antiparticles, like antielectrons, was assumed. This seemed quite radical, but flexibility was a hallmark of the Bohr group. At Pauli's suggestion, they had already assumed the existence of a neutral particle that was sometimes paired with the electron called the "neutrino." (Named for reasons that will be explained shortly.) Likewise, Bohr's group was sympathetic to the ideas of a Japanese physicist, Hideki Yukawa, concerning a light neutral particle called a "meson" whose job it was to hold the protons in the nucleus together. Certainly, the idea of matter with opposing charges was well accepted: the electron was clearly proven to be negative and the proton was clearly proven to be positive, but proofs of the existence of Pauli's and Yukawa's neutral matter and Dirac's antimatter would await experimental verification outside the Bohr group.

### Einstein Insists on Overall Conformity of Physical Laws for Both Tiny Particles and Giant Galaxies

Curiously, even as Einstein's views were being accommodated by Dirac—and even as Einstein was becoming something of a world-wide celebrity—Einstein himself grew further away from seeing the quantum mechanical view of the atom as satisfactory. In a manner that was ironically reminiscent of the archly conservative Planck, Einstein, the iconoclast, felt that the approach of making endless adjustments to the Bohr model was very untidy. He found the many separate announcements of new particles and new forms of radiation to be like the study of individual trees, when what was needed was a look at the whole forest. After the 1930s, Einstein devoted himself more and more to what has now become known as Grand Unified Theory (GUT) or generalized field theory. Followers of field theory did have one advantage over straightforward particle experimenters. They continued to draw on astrophysical findings, much as Einstein had done, to demonstrate the viability of their views. This has resulted in a strange but continuing connection between the world of the galactically large and the infinitesimally small in journals that cover particle physics, field theory, and astrophysics. A good way to sort out the true field theorists from physicists who consider relativistic effects to be one of many factors to be taken into account is to assess how often notions of the creation of the universe and the geometry of space enter into the discussion of a given particle's behavior. The more this is the case, the more likely it is that one is dealing with a field theorist.

## Splitting the Nucleus Leads to the Discovery of the Powerful Neutron

The phenomenon of radioactivity did not go away quietly, even though academic interest and progress came much more rapidly in theoretical work. At German universities, Walter Bothe and Herbert Becker began to notice that certain radioactive rays were able to routinely knock protons out of the hard little nucleus of some atoms. This was better than even Rutherford was able to regularly accomplish. Rutherford's positively charged alpha particles were usually strongly repulsed from the positively charged nuclei before they could do much damage. What new particle or wave could be so powerful?

The work of Bothe and Becker was extensively furthered by Madame Curie's daughter and son-in-law, Irène and Frédéric Joliot-Curie. They found that the target atoms for these new beams became so unbalanced by the collision that the targets themselves became radioactive for a short while and gave off beta rays useful for medical diagnosis (a discovery for which they won the Nobel Prize in 1935). In time it was realized that whenever the new beam directly knocked out some positively charged protons, an exactly offsetting number of negatively charged electrons had to be ejected through beta rays until the atom was once again electrically neutral.

A British theoretical physicist, James Chadwick, suggested that the powerful rays that were doing the damage could not be merely electromagnetic energy in waves. Indeed, it was soon conclusively shown that these rays or waves could not be deflected by conventional electrical fields or magnetic targets. They had to be a straightforward beam of heavy particles of no electrical charge. Today we call these neutral particles "neutrons." In order to distinguish these clearly demonstrated heavy neutral particles from the lightweight neutral particles theorized by Pauli, an Italian wag named the lighter particles "neutrinos" (baby neutrons!)

### *Fermi Tames the Neutron — for a Time*

The Italian with the sense of humor was Enrico Fermi, perhaps the greatest scientist Italy has produced since Galileo. He was the son of a self-educated railway worker, and was himself a voracious reader many years ahead of his chronological classmates. Despite his extreme youth, Fermi rose swiftly in the ancient and formidably hierarchical world of Italian universities. He was a full professor before the age of thirty. Although he was actually liberal in his personal political philosophy, and was publically noncommittal, he received a good deal of sponsorship from the fascist government. Mussolini saw in him a chance for fascist Italy to achieve international respectability within the sciences. Fermi came to quantum mechanics armed with excellent mathematical insights — one of the few strengths of the old Italian university tradition. He became one of the first scientists outside of Bohr's circle to make substantial theoretical contributions to the field, dividing his papers between the internationally recognized *Zeitschrift fuer Physik* and the more local *Nuovo Cimento*. (Indeed, his papers were to promote *Nuovo Cimento* to the rank of an international organ.)

Fermi took up experimental work only when in his mid-thirties (comparatively late in his career)! Intrigued by the flurry of reports concerning the power of neutrons, he assembled a team of the best among young Italian physicists, most notably including Emilio Segrè. Like Bohr, Fermi socialized extensively with his group and, indeed, members characterized each other by irreverent nicknames. (Fermi was "the Pope," Segrè was "the Cardinal.") Together they rigged up their own neutron sources and targets, even to the point of building their own Geiger counters. Fermi hit upon the idea that many more unstable "artificially radioactive" compounds than the Joliot-Curies had made could be created if these powerful neutrons could be slowed down a little. He correctly theorized that more disruptions of normal, stable nuclei would occur if the neutron did not just shoot clean through the nucleus. Each of the high speed neutrons of the Joliot-Curies tended to knock out only a single proton in its rapid transit, yielding only a single offsetting electron ejection. Rather, Fermi wanted each disruptive neutron to stay awhile in the nucleus and tumble around, wreaking a kind of nuclear indigestion, bumping off multiple protons with a longer-lasting ejection of electrons afterwards. He succeeded by developing a list of substances that slowed down the neutrons before they hit their target. (Indeed, one day the light-hearted experimenters stole the fish tank of a curmudgeonly elder professor, and found prophetically that the water — presumably including the goldfish — served as a pretty fair modifier of neutron speed.) The slowed-down neutrons were taken in large numbers by the unsuspecting target nuclei and the subsequent radioactive belching was considerable and longer-lasting. As he had predicted, Fermi was able to generate a wider and much more powerful assortment of artificially radioactive versions of target elements.

## War Clouds Gather and Nuclear Research Becomes Secret

The fun would quickly go out of Fermi's finding, for a short time later, German scientists Otto Hahn and Liese Meitner discovered that neutron bullets could entirely break up heavy nuclei, such as that of uranium. This nuclear reaction yielded a great deal of energy, and something more than the expected mild-mannered proton and electron ejections. From every shattered uranium nucleus came a spray of devastating neutrons to multiply the effect with lightning-like speed. Many scientists worldwide and a few European military officials quickly sized up the prospects of this discovery with the idea of a tremendously explosive weapon down the road — but only if enough of the right kinds of radioactive material could be prepared. The right kind of uranium, uranium 235, was very scarce and extremely hard to refine. Virtually everyone decided that it was better to increase the radioactivity of other ultraheavy metals by Fermi's method than to try to scrape together the right amounts of difficult and scarce uranium.

Ironically, it was discovered that a special type of water, called "heavy" water because it contained an extra neutron that made it naturally heavy, was an excellent moderator of neutron speed. (Recall Fermi's fish tank experiment!) Many a military mission would focus on destroying the enemy's heavy water isolating facilities in the coming war. The international scientific community was forced in many cases to choose sides between the Allied and Axis scientific armies.

Einstein, who had already emigrated to the U.S., petitioned Franklin Roosevelt to start up U.S. efforts. (Interestingly, American military officers were highly skeptical and had little respect for fuzzy academic types like Einstein.) Bohr was smuggled out of Nazi hands in a British fighter plane. (The FBI insisted that he maintain a false identity during his extended stays in the U.S. However, the man was as famous in scientific circles as Einstein and only slightly less striking in appearance, given his height and strong features, and the ruse never worked!)

Fermi used his trip to Stockholm to accept his Nobel Prize in 1938 as an opportunity to defect to the U.S. Segrè and Pauli had quietly preceded him. Revealingly, Mussolini did not grasp that Fermi was really defecting. Mussolini thought that Fermi was merely respecting a temporary banishment decree issued by the Duce. Mussolini was personally hurt that Fermi refused to wear the Mussolini-look-alike uniform to the Nobel Prize awards ceremony that he had sent for the purpose! (Fermi wore the customary white tie and tails instead.) Schrödinger sat out the war in neutral Ireland. Meitner escaped to neutral Sweden.

Stark and Lenard, however, enthusiastically embraced the Nazis. Heisenberg cooperated with the German military despite an evident personal distaste for Nazi upstarts. There is now considerable evidence that he deliberately sidetracked German bomb research. (He met with Bohr secretly shortly before Bohr's defection.) However, given Heisenberg's immense prestige (German military officers, in contrast to their American counterparts, tended to accord gentile Nobel Prize-winning university professors an almost exaggerated respect) he was never contradicted.

### Fermi in America

One of Fermi's most distinguished accomplishments came after his arrival to the U.S. but just before the war. It was at this time that he developed his notions of controlling the chain reaction effect in radioactive materials by his ability to find neutron-slowing and neutron-capturing materials. By interspersing neutron moderating materials among highly radioactive materials he was able to raise and lower the energy levels of a "radioactive pile" as early primitive reactors were unceremoniously known. His first installation was built under the disused grandstands of the football field at the University of Chicago. (The cerebral University of Chicago had dumped football by then.) With modifications, Fermi's type of reactor would become the standard for even today's civilian use. Ever affable and continuously productive, Fermi would become the dean of U.S. nuclear studies after the war until his premature death, which was probably due to years of radioactive exposure. More pertinent to the wartime years, by-products of Fermi's peaceful nuclear pile included plutonium, an artificial element of enhanced radioactive potential, created within nuclear piles by the absorption of heavy particles by the nuclei of uranium 238, a much more common radioactive material. Frighteningly, this waste product of peaceful nuclear energy generation served as an excellent substitute for the much rarer uranium 235 in early atom bombs.

## Oppenheimer and the Fission Bomb Team

The U.S. atom bomb project was headed by J. Robert Oppenheimer. He was a distinguished Berkeley particle theorist who was effectively drafted to head up the A-bomb team in remote New Mexico. His scientific staff included some of the most talented young Americans as well as the cream of Continental nuclear physics. Many among the European contingent were Jewish refugees, most of whom were also Socialists fleeing persecution at the hands of the Nazis. The brilliant Oppenheimer was chosen in part because of his familiarity with them on a professional basis since Oppenheimer, like many of these scientists, was yet another alumnus of stays with Bohr. Moreover, it was thought that Oppenheimer would be empathetic since he was Jewish by birth, if not by observance, and had a long and public involvement with Socialist and Communist activities within the U.S.

Oppenheimer had a substantial change of heart concerning his weapons work once it was seen just how devastating the atom bomb would be. He was joined by many on his team in his doubts and demands for a more reasoned approach to nuclear armaments. In later years their concerns would be voiced by a journal of strong political/ethical contents whose title — *Bulletin of the Atomic Scientists* — bears witness to its origins among this historic group. Oppenheimer would later clash with defense authorities on both the wisdom and the technical feasibility of the hydrogen or fusion bomb. His highly open Communist sympathies, which were tolerated by military authorities during World War II for the sake of his unique mix of qualifications, were now regarded as a tremendous risk. The issue became particularly acute when Oppenheimer suggested that if the U.S. held back on developing further weapons, the Soviets would probably stop developing them as well, leading to a safer world. Seen as an unjustly persecuted man, Oppenheimer became something of a folk hero to academia. By sharp contrast, one of Oppenheimer's old bomb-making teammates, Edward Teller, a scientific equal and ideological opposite, would be cast as the promilitary devil incarnate.

## Kurchatov and the Rise of Soviet Nuclear Physics

While both the modest German program and the even smaller Japanese atom bomb effort died at the end of the war, the Soviet effort that had begun during the chaos of the Nazi invasion of Russia expanded. It was to succeed only a few years after the war in producing conventional "fission" atom bombs, under the leadership of Igor Kurchatov, the dean of nuclear physics in the Socialist Bloc. While much more formal in personality than Fermi, his career mirrored the transplanted Italian's in both positions and honors. Like Fermi, he was to die prematurely from cancer, probably due to years of overexposure to radioactive materials. Even his legacy was like Fermi's, for his assistants included future Nobel Prize winners like Igor Tamm and Andrei Sakharov. For a number of decades, Soviet journals of nuclear physics were carefully scanned in every serious scientific or military collection in the West, given the continuous advances made by the Russians.

Scientifically, a tremendous amount about nuclear chemistry and radiation effects were learned in those Cold War years, and some journals covering these subjects are discussed in the chapter on inorganic chemistry.

## The Era of the Big Machine: Accelerators Speed the Progress of Peaceful Particle Physics

The theme of collisions disclosing new nuclear properties or particles had become constant by the 1930s. Most of these collisions, however, were at the mercy of unpredictable emissions of rays or particles from largely natural radioactive sources. Not only was there some uncertainty of the rate of emission of these projectile particles but their energy or range of speed upward could not be accelerated. (Fermi could slow down particles but not accelerate them beyond their initial ejection speed.) The idea of more control over nuclear and particle experiments was ripening.

Since Geiger's time, moreover, skill had been built up in the number and variety of radiation and particle detectors. It occurred to a small group of pioneers that if electrically charged particles could be injected into a near vacuum (to minimize unwanted collisions of particles with air molecules that would slow things down) repeated pulsings of electrical charges could speed particles along. The idea was tried and worked very well. Further, by using electromagnetic fields to steer the particle stream, particles that had accelerated to sufficient speeds might then be smashed into heavy nuclei to create new artificial elements or, as became increasingly common over time, to discover new particles in the broken-off fragments as their tracks showed up on the new detectors.

In Britain, the effort was headed by Sir John Cockcroft and Ernest Walton. In the early years they used protons (hydrogen atoms minus their electrons) to bombard light elements like lithium, and create short-lived radioactive versions of light compounds. Over time they used neutrons and heavier nuclei to bombard uranium to create new superheavy elements. They won the Nobel Prize in 1951.

In the U.S., the effort was headed by Ernest O. Lawrence, a graduate of Minnesota and Yale who worked largely at Berkeley. He invented the cyclotron type of accelerator in which the particles are spun around along a circular racecourse, attaining tremendous speeds before being veered off for collisions. Most research accelerators today bear traces of this fundamental design. Lawrence's collaborators and achievements were numerous, and he was one of the first peacetime users of the large multiskilled team concept in his accelerator labs. He involved graduate students, post-docs, theoretical colleagues within physics, electrical engineers, and an army of machinists, electricians, plumbers and technicians. His historically most important protégé was J. Robert Oppenheimer of the atom bomb effort. Other important associates included a chemist, Glenn Seaborg (Nobel Prize, 1951), who produced seven new elements and forced a rearrangement of the periodic table, and Emilio Segrè, (Nobel Prize, 1959) Fermi's former partner in crimes against fish-tanks. Indeed, Segrè, who himself had created new superheavy elements and became a leader at Berkeley after Lawrence's retirement, decisively turned the emphasis in accelerator work away from creating these ever-heavier elements towards finding new particles from collisional studies.

Over time, Lawrence's style of large group collaboration, which started before the war, was reinforced by the large group effort of the atomic bomb work. Even though nuclear research shifted out of the immediate control of the military after the war, many of the civilian advisors and grants administrators carried over the large group style. Moreover, the machinery of accelerators and particle detectors demanded more and more staffing and  money.

### Sorting Out the Major Energy Ranges and Purposes of Today's Accelerators

Accelerators, which were initially the size of washing machines, grew to the size of classrooms, then gymnasiums, then stadiums, then campuses, then small towns, and are now proposed to grow even larger. Today, all around the world there are what amount to small cities of scientists and technicians who run these increasingly complicated and millions-of-times-more-powerful accelerators. In the U.S., the largest operational accelerators are at the Fermi National Accelerator in Batavia, Illinois (a machine about four miles around with two rings that eventually lead opposing beams into collisions), and at Stanford (twin two-mile straight-line tunnels that abruptly curve in toward each other for collisions).

In Europe, refugees from Nazi Germany like I. I. Rabi, a member of Oppenheimer's World War II team in America, encouraged the development of a purely scientific, purely peaceful multinational European facility to compete with the Americans. The response has been amazingly strong and positive. Today, fourteen European governments have built an elaborate complex in Switzerland called "CERN: Conseil Européène de Récherche Nucleaire" — the title is French but everyone publishes in English. CERN's biggest machine is a whopping seventeen miles around. Not to be outdone, the U.S. is proposing to build a $7 billion, fifty-four-mile circumference machine in Texas that will use superconducting technology to reach forty trillion electron-volts, a billion times more powerful than Lawrence had at his disposal.

There are, moreover, at least a dozen other significant accelerator facilties in the U.S., usually run by consortia of universities with federal support. There are a number of older but still potent machines in individual European countries and in Canada. Japan has impressive, fairly new, medium-sized machines in their Tsukuba Science City. Virtually all of the machines mentioned thus far are termed "high-energy" and are used for fundamental particle hunting — a distinction from "medium"-and "low"-energy machines that shifts back and forth as newer machines are built or older ones improved. India, the People's Republic of China, and the Third World have some high-energy machines, but more typically work at low or medium energies, and generally deal with better understood composite particles. These countries also contribute equipment and frequently send collaborators for multinational efforts. Russia, formerly a stand-alone competitor in high-energy physics, has not kept up in the last decade. It now involves itself primarily in CERN projects when seeking high energies and more fundamental particles. The Soviets continue working at medium and low energies at their own, older facilities back home.

As had been mentioned in the beginning of the chapter, certain lower-energy machines and the larger particles they work with, have continuing scientific interest and productivity. Indeed, although it sounds confusing, low- and medium- energy machines are finding new uses as heavy ion accelerators. There has been a good deal of ingenuity shown in the U.S. and abroad in tailoring older machines for new slants in larger, composite particle work. These rebuilt machines, many of which are taking advantage of opportunities for downsizing through electronic miniaturizations, are far more practical for a wide range of studies involving fields like surface science (see the applied physics chapter) and for the production of medical isotopes and antitumor therapeutic radiation.

A political question remains: what has been the end result of all the highest-energy expenditures? The answer is about twenty Nobel Prizes and an annual issue of *Physics Letters* that runs over a hundred pages of new information on particle properties and behaviors.

Still, the picture of the properties and behaviors of particles obtained from accelerators remains incomplete without a consideration of wave and field theory. Most of this quiet reflection on waves and fields comes from outside the giant communities of accelerator scientists.

### The Accelerator Moguls Sustain the Cottage Industry of Theoretical Particle Studies and Field Theory

If, because of costs and the enormous commitment of land and electrical power, there are only a few dozen significant centers of particle experimentation worldwide, there are hundreds of colleges, universities, and research institutes that employ theoretical physicists who are interested in particles and fields. The qualifications for membership in this latter group are largely advanced abilities in mathematical conceptualization and a good library. To a substantial degree, theoreticians continue Einstein's approach of looking for a consistent, simplified theory. Like Einstein, these theorists expect to find confirmation of their theories not only in the findings from accelerator studies but in astrophysical phenomena and in creation-of-the-universe scenarios ("cosmogonies"). Even in those cases where the theories fail their grand mission of explaining everything, they often explain a large category of particle behaviors. Indeed, some of the mathematical findings turn out to have transferability to other, more applicable domains of physics. At the very least, it has largely fallen upon theoreticians to catalog, compare, and update the growing inventory of experimental facts about particles and cosmic forces for subsequent "meta-analysis."

To a substantial degree, theoretical physicists set the long-term agenda of what is sought by experimentalists. Experimentalists tend to gain fame for two achievements: betting on the right theory and then finding the right particle. As in horse racing, long shots pay off the most spectacularly. Dirac's 1930 predictions of antimatter were most decisively confirmed by the finding of antiprotons by Segrè and Chamberlain in 1954 (Nobel Prize, 1959). Yukawa's meson of 1935 was definitively detected only in 1947, winning Cecil Powell the Nobel Prize in 1950. Pauli's 1931 neutrino were not found in cosmic rays until 1956, and were not found through particle collisions until Ledermann, Steinberg, and Schwartz succeeded in the 1960s (winning a Nobel Prize for the latter group in the 1980s). Indeed, much that is known or conjectured concerning the fundamental particles that carry forces and energy fields—particles today characterized as "bosons"—has been worked out well in advance by theoreticians. And not all of the theoreticians need to come from instrumentally rich countries: "bosons" are named in honor of an Indian physicist, S. N. Bose.

Finally, it has fallen to theoreticians to explain how atoms and particles stay together and function on an ordinary basis. Apart from the moment of the creation of the universe, and the violent inner workings of stars, most matter is not being created, spectacularly altered, or shattered by being slammed into a "brick wall" at something approaching the speed of light. Much remains to be understood in the gentle manner of Yukawa about why nuclei are stable (apart from radioactive decay) and why particles that we now know are composite generally function in one piece, or at worst, transform without too much drama. A smaller number of Nobel Prizes have been awarded in more recent years for convincing explanations of relative normalcy, as it were. These studies of interesting or ironic situations within the atom do not necessarily require a small city of scientists to uncover them, just some very clever ones.

Certainly the most popular, colorful, and influential explicator of particle physics since Fermi has been the late Richard Feynmann of Caltech, a 1965 Nobel laureate. If there is a given author to be recommended for enriching a layman's understanding of particle physics and the role of a genuinely humorous and good-natured man in that field's development, it is Feynmann. Anyone whose best-selling titles are *Surely You're Joking, Mr. Feynmann!* and *What Do You Care What Other People Think!* certainly piques this author's interest.

Yet even as mathematically inclined theoreticians have greatly helped find particles and explain fields, a growing number of them are looking at a fundamentally different approach: supersymmetric theory. While some experimentalists decry its development as yet another thought trend not based in reality—much like the gauge theory that was the rage of the 1980s—supersymmetric theory is likely to generate considerable interest within existing journals and perhaps spawn its own titles in the 1990s. The fundamental unit in supersymmetric theory is not a particle or field but rather a mathematical construct, existing in ten dimensions, called a "superstring." In superstring theory, all the differing particles are actually manifestations of the same fundamental superstring vibrating in differing ways in given dimensions. Superstrings would represent something of an ultimate grand unification, but even so, it is arguable whether or not superstrings could explain gravitational fields across distances, yet another Einsteinian test of theoretical consistency. Indeed, the problems of gravitational attraction must be solved and the events at the creation of the universe must be incorporated in any theory that would satisfy the strictest unification theorist.

### Plasmas on Earth and in the Stars

One of the side effects of the work on accelerators has been the rise of a related field called plasma physics. Recall that accelerator physicists needed to get electrically charged particles in order to electrically accelerate them. It was found with time that intensely heating gases would often do the trick, because at extreme temperatures, electrons are stripped away, giving an overall positive charge to what remained of the atom. Interestingly the "plasmas" that result follow some of the laws of ordinary gases and liquids—an area called fluid mechanics—and some of the laws of electromagnetic waves, because of the electrical charges involved. Most pertinently, plasmas also have some peculiar physical laws of their own!

Plasmas became theoretically interesting when it was suggested (first by Cambridge University astronomer Sir Arthur Eddington, and then by his most famous student, Chandrasekhar, an Indian who emigrated first to Cambridge University, then to the University of Chicago) that the transforming fusion of lighter plasmas like hydrogen into heavier plasmas of helium might account for the tremendous output of energy of the stars. Astrophysical studies had definitely shown the presence of these plasmas in the spectra of stars, and astrophysical calculations had suggested that even the enormous energy of nuclear fission was insufficient to account for all the energy radiated by a star.

In Russia this theme was picked up by Igor Tamm, one of Kurchatov's protégés and a member of the Soviet bomb team. Tamm himself had a protege in the young Andrei Sakharov. These two had excellent exposure to both the longstanding scientific interest in gas physics and fluid mechanics, and to the latest developments in nuclear physics. Sakharov in particular seized the idea that it might be possible to build a hydrogen-fusing-into-helium "H bomb" even more powerful than the fission "A" bombs then currently being investigated by Oppenheimer and Kurchatov. He had his doubts about the wisdom of the project but was relatively confident of overcoming the technical problems. Indeed, he quickly perceived that his supervisors were bent on pursuing the project with or without his participation. Sakharov succeeded beyond expectations and the Soviet Union followed the American success by a few months, when it was thought it would take years for the Soviets to catch up. Curiously in Sakharov's forthcoming memoirs, which have been excerpted in a number of popular magazines, J. Robert Oppenheimer was portrayed as highly principled but both technically incorrect and politically naive in the extreme in his resistance to the H-bomb effort. Edward Teller, perhaps the most academically vilified scientist of the twentieth century for his pushing of the H-bomb project in the U.S., and for his denouncement of Oppenheimer, is praised by Sakharov as equally principled as Oppenheimer, technically more farseeing than Oppenheimer, and tremendously more astute about the nature of Soviet intentions in those days than was either Oppenheimer or Sakharov himself. When the ailing Sakharov was finally allowed to travel to the West, he was asked whom he most ardently wished to meet — noted pacifists? other Nobel laureates? leftist academics? Sakharov instantly replied that he wished to see Teller, and to be seen by all the others as he embraced this maligned "Father of Darkness."

The Soviets have maintained an exceptional interest in the peaceful uses of energy derived from plasma fusion. (Tamm was to win a Nobel Prize along with Cherenkov for peaceful nuclear physics in 1958, and Sakharov took the peace prize in 1988.) Using the concepts of workers such as Sakharov, the Soviets have constructed a number of fusion chambers in which very hot plasmas are further heated and compressed by powerful magnetic fields in an attempt to simulate and understand fusion processes. One practical goal is to attain the safe and economical production of energy from plasma fusion. The machines, called Tokamaks, from a Russian language acronym, are abundant in the Soviet Union, and a number of them have been constructed for U.S. research efforts, largely at Princeton. (There are even newer machines in Japan and West Germany.)

Other U.S. efforts have focused on such novel strategies as superheating and compressing tiny pellets containing heavy hydrogen with jolts from very powerful lasers. To date, the energy released from any fusion that results has not matched the tremendous energy expended in providing these laser blasts in the first place. Most researchers do not expect to see economical energy production from "hot fusion" for many years to come, although there have been and will be many scientific and technical dividends along the way. While there has been a faint hope that "cold fusion" would some day come to fruition, a more practical earthbound application of contemporary plasma work has been the production of novel superthin coatings, a subject for the next chapter.

Astrophysicists continue to observe and study plasma behavior and thermonuclear physics on a daily basis. In fact, journals of astrophysics are frequent sources of ideas for earthbound experimental plasma physics workers in the same way that journals of theoretical particle and field work are idea sources for accelerator scientists. Plasma science has provided yet another reason for the interlocking of interests of atomic, nuclear, particle, and field theory workers with astrophysicists.

## SORTING OUT THE JOURNALS

### Journals of the Low Energy Physics of Intact Atoms

Journals which primarily deal with small aggregates of atoms, whole atoms, or the electron shells of intact atoms are relatively few in number. (See Figures 1 and 2.) This paucity is not because atomic physics is out of fashion. Rather, some of its topics are being poached by the journals of neighboring disciplines. Journals of chemical physics have reached into the area of fine level spectroscopy, as have some journals of atomic spectroscopy from the optics community. Nonetheless, published studies of the structure,

FIGURE 1

## Contributors and Relative Impact Leading Atomic Physics Journals

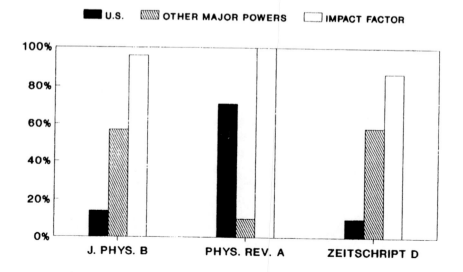

FIGURE 2

## Comparison of Number of Papers, Costs, and U.S. Market Penetration

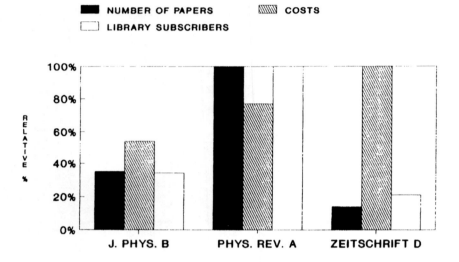

vibration, and lower energy collision behavior of this level of matter still number over a thousand annually.

The first choice in this area is *Physical Review A*. This journal has an interesting history in being the "catch-all" section of the alphabetical *Physical Review* series. The "catch" has historically included papers in atomic physics. But even as we write, this American Physical Society journal is undergoing a fission of sorts, with differing subtitles for each of its semimonthly issues. Issues at the first of the month, informally called "A-1"s (like the steak sauce), stress more general atomic, molecular, and fine-structure optical matters. Issues at midmonth, the "A-15"s, stress thermodynamics, plasmas, and certain interdisciplinary subjects. Despite its eclectic nature, *Physical Review A* is the leader in virtually all measures affecting library selection for atomic work.

Since both competitors for second choice have similar impact factors and similar overall proportions of papers from scientifically advanced countries, a preference of one over the other may stress differences in scientific ethnicity and costs. The best British and Commonwealth papers in this field come from the *Journal of Physics B - Atomic, Molecular and Optical*, an Institute of Physics publication. Many of the better Continental papers come from the Springer's *Zeitschrift fur Physik D - Atoms, Molecules and Clusters*. While virtually any program with a serious interest in atomic-level physics should have both titles, costs greatly favor the British title if only one can be taken.

Other journals which are not quite exclusively devoted to atomic level physics that might be considered are the *Canadian Journal of Physics* from the National Research Council, and *Physica Scripta* from the Swedish Royal Academy on behalf of a consortium of Scandinavian physical societies. Both journals continue to reflect the traditions of historical leaders in atomic physics, especially the spectroscopic traditions of Herzberg in Canada (recall the previous chapter) and the Siegbahns in Sweden.

One title that is not quite comparable to a typical journal of atomic physics, but which ought to be taken in any collection with a serious emphasis on experimental work, is Academic's *Atomic Data and Nuclear Data Tables*. This serial is effectively an ongoing reference and review collection featuring very lengthy tables with experimental details and calculations.

There are a number of subspecialty and neighboring specialty journals that will also be required in some collections. For example, strongly spectroscopic programs will want the *Journal of Electron Spectroscopy* from Elsevier, and the *Journal of Quantitative Spectroscopy and Radiative Transfer* from Pergamon. Likewise, spectroscopists will need the *Journal of the Optical Society of America B - Optical Physics*, where some atomic fine-structure is probed.

Virtually every collection will also need the *Journal of Chemical Physics*, the dominant journal of molecular matter discussed in the physical chemistry chapter.

The best review in this area comes from Academic: the hardbound series *Advances in Atomic, Molecular and Optical Physics*.

### *Journals of Mid-Energy, Heavy-Ion Nuclear Studies and Accelerator Technology*

The journals of this assortment feature studies of nuclear structure, middle-range collisional reactions between nuclei, particularly "heavy ions," natural and induced radioactivity, and the better-understood nuclear particles. This area of studies produces about twice as many experimental papers as either the larger atomic scale or the smaller subnuclear scale of matter. This is in part because the experimental equipment involves less cost and is more widespread, and because some beam generating work is now becoming applicable to other areas of science.

Figures 3 and 4 illustrate today's leaders at approximately this level of energy and scale of matter. The qualifications for this roster of journals will always be a little fuzzy in that some of these journals also take some higher-energy papers (a good example is *Jour-*

FIGURE 3

## Contributors and Relative Impact
## Leading Nuclear Physics Journals

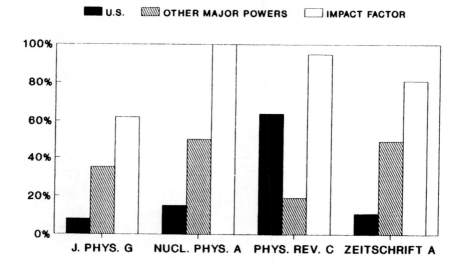

■ U.S.    ▨ OTHER MAJOR POWERS    ☐ IMPACT FACTOR

J. PHYS. G    NUCL. PHYS. A    PHYS. REV. C    ZEITSCHRIFT A

FIGURE 4

## Comparison of Number of Papers, Costs, and U.S. Market Penetration

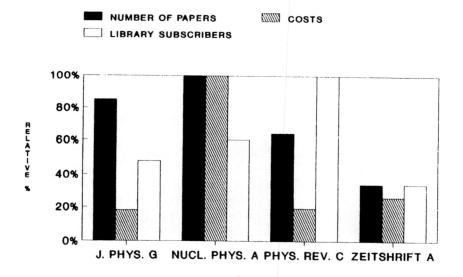

■ NUMBER OF PAPERS     ▨ COSTS
☐ LIBRARY SUBSCRIBERS

*nal of Physics G - Nuclear and Particle*) while some fairly low-energy papers will appear in *Nuclear Physics A*.

The usual assumption, that the American society journal, in this case *Physical Review C - Nuclear Physics*, is automatically the most important journal and should be the first choice may not be true. This American Institute of Physics title is, in fact, strongly pressed by *Nuclear Physics A* from Elsevier. The Elsevier title would have an even clearer lead if Springer's *Zeitschrift fuer Physik A - Atomic Nuclei* was not taking away some of the other leading continental European papers. Nonetheless, *Physical Review C* does have an excellent mix of papers, competitive impact factors, and, if only on costs, wins out in more American libraries than any other title in this group. Given any extra money, however, *Nuclear Physics A* is clearly a necessity, and the *Zeitschrift* only slightly behind.

Curiously, the *Journal of Physics G* does not fare as well as might be expected for a British society title. Part of this is due to a lack of heavy spending in indigenous British nuclear and particle machinery. And part of this may be due to the fact that Britain's current reputation in nuclear work is probably based more on the U.K. as yet another CERN partner publishing most of its best work in a CERN-oriented journal like *Nuclear Physics A*. Moreover, the British *Journal* probably publishes more Third World and Commonwealth papers than any other member of this comparison group, and this reduces its cachet somewhat. However, the *Soviet Journal of Nuclear Physics*, translated by the American Institute of Physics, is a title that nonetheless ranks below the *Journal of Physics G*, and for some of the same equipment reasons. Moreover, few non-Soviet papers appear there. It should be taken only after the British entry is well-secured in the budget.

A pair of Elsevier journals for experimental nuclear physics stresses the enormous amount of equipment for nuclear probing and particle hunting. The *Nuclear Instruments and Methods in Physics Research*, A and B series, is edited by Swedish Nobel laureate Kai Siegbahn. *Section A* stresses plasma generators, vacuum chambers, detectors, steering magnets, target construction, etc. *Section B* is devoted to the applications of particle beams to the characterization and tailoring of surfaces, one of the first applications of nuclear work to day-to-day science. This will be expanded upon in the chapter on applicable physics.

These Elsevier accelerator titles do see some competition from the IEEE, the U.S. society for electrical engineering. Recalling the enormous size and multidisciplinary nature of accelerator teams, it should come as no surprise that highly competent technical support is necessary, and that these support personnel need a journal. The *IEEE Transactions on Nuclear Science* serves this cadre, while also dealing with some problems of more pragmatic nuclear physics, including radiation detection, medical use of particle beams, and nuclear power supply production. The price for this IEEE title is a small fraction of its capable Elsevier competition. Nonetheless, both the Elsevier series and this IEEE title should be in every accelerator facility collection.

The best overview in the overall area of nuclear physics is the *Annual Review of Nuclear and Particle Science* from the nonprofit

Annual Reviews, Inc. A more cosmopolitan group of contributors will be found, however, in Pergamon's *Progress in Nuclear and Particle Physics*, and is recommended for larger collections.

## Journals of High-Energy Nuclear Physics and Both Theoretical and Experimental Particle Hunting

Figures 5 and 6 portray those journals specifically devoted to the highest energy nuclear physics. These titles stress some experimental work and a good deal of theoretical interpretation, especially involving fields and cosmic rays. The reason for the limited amount of direct experimental reports is that four nominally general interest physics journals eagerly seize announcements of new particle findings.

Three of these journals are as follows: *Physical Review Letters* from the American Institute of Physics, *Europhysics Letters* from a

FIGURE 5

# Contributors and Relative Impact Leading Particle Physics Journals

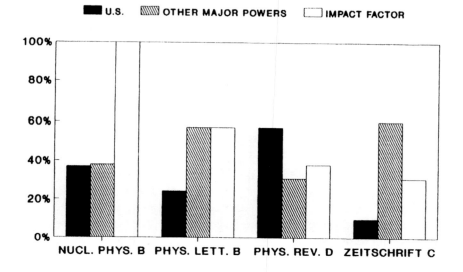

FIGURE 6

## Comparison of Number of Papers, Costs, and U.S. Market Penetration

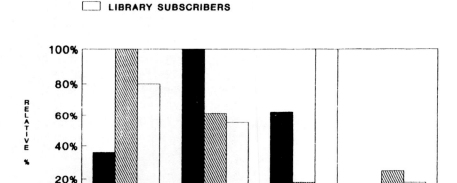

Franco-Italian consortium, and *JETP Letters*, a Plenum Press translation of the letters section of the Russian language *Journal of Experimental and Theoretical Physics*. All three publish some papers in other subjects, but must absolutely be kept on a par in collection priorities with those titles more obviously dedicated to particle and field work. A fourth title, *Physics Letters B*, from Elsevier, has become virtually all particle and field work, with a little general nuclear physics thrown in. It rightfully must be compared explicitly with this particle physics journal group.

If *Nuclear Physics A* was a strong contender at more moderate accelerator energies, *Nuclear Physics B* is frankly dominant in journals which mix the highest energies with theoretical work. It is unheard of that almost 40% of a large foreign journal would be U.S. papers when there is a much less expensive, reputable, U.S. society competitor, but this is exactly the case in this group. *Physical Review D - Particles and Fields* from the American Institute of Physics is by no means of poor quality, but it has simply not attracted the leading theoreticians in enough numbers to offset for the loss of sexy new particle verification experiments to its companion *Physical Review Letters*. In particular (pun intended) *Physical Review D* has not become required reading for the persistent number of mathematics PhD's who make important advances to particle field theory. By contrast, *Nuclear Physics B* and *Physics Letters B* have a healthy relationship. It is interesting to note that the strongest journal of mathematical physics, *Communications on Mathematical Physics* from Springer, is typically one of the top five journals cited by authors in *Nuclear Physics B* and *Physics Letters B*.

To some degree, the Springer entry, *Zeitschrift fuer Physik C - Particles and Fields*, seems left out in the cold in this otherwise hot assortment. Yet neither the German effort nor this German-based journal is an also-ran. While the prosperous Germans have become quite involved in CERN facilities in Switzerland they are the leaders in individual national facilties in Europe as well, and sponsor international collaborations on their own soil. Indeed, in recent years it can be safely ventured that they are second only to the U.S., and have not yet been surpassed by the Japanese.

The *Zeitschrift,* moreover, is a tremendous historical survivor. Its three-way split to serve the differing scales-of-matter was an example of enduring competitive spirit and flexibility. The *Annalen der Physik*, its old rival, fell into East German hands and has virtually lost all its old glory. Most particle and field physicists today think of Academic's *Annals of Physics* when they see *Ann. Phys.* as a reference, and this latter day *Ann.* is especially commendable for its lengthier, more speculative papers.

A final historical note concerns *Nuovo Cimento*, a title connected with the remarkable achievement of people with Italian roots in nuclear and particle work. While its ongoing *Section B* is a title of moderate urgency in the particle and field world even now, its old "letters" section, *Lettere al Nuovo Cimento*, has merged with the letters section of the French *Journal de Physique*, to give us the critical journal *Europhysics Letters*, mentioned above.

### Journals of Field Theory

The journals in this grouping place a very heavy emphasis on field theory, particularly gravitational theories, in any discussion of particle existence and function. Cosmological concerns and findings from astrophysics are quite common. Mathematically inclined authors are dominant, and few experimental papers per se are accepted. Once again these titles are just as likely to cite journals of mathematics as they are of physics, with *Communications in Mathematical Physics* being the prime example.

To a substantial degree, these are the journals of Einstein's heirs in grand unified theory. Indeed, lengthy, almost classically reflective papers are found in a number of these titles. (See Figures 7 and 8.)

Plenum Press, a high-quality, American, for-profit firm, publishes three of these titles, and a fourth is just getting under way. *Foundation of Physics* is the most historical/biographical/philo-

FIGURE 7

## Contributors and Relative Impact
## Leading Field Theory Journals

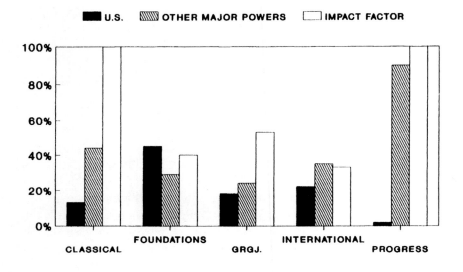

FIGURE 8

## Comparison of Number of Papers, Costs, and U.S. Market Penetration

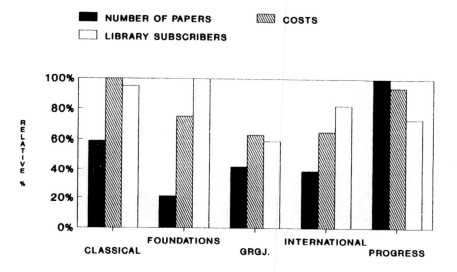

sophical among all these titles. It tends to treat this area of physics as strict Freudians treat psychotherapy: by continuous referral to the founding fathers. Nonetheless, *Foundations* succeeds to a far greater degree than does its rigid counterpart in psychiatry, so much so that *Foundations of Physics Letters* (not shown in the figures) has recently commenced as a companion journal for shorter papers. *General Relativity and Gravitation* is a related publication from the International Committee on Relativity and Gravitation with production and distribution handled by Plenum. It takes mostly field papers and has a definite cosmology outlook. Its fewer particle papers understandably include some discussion of gravitons (particles which some theorists feel are responsible for gravitational attraction) as well as particle behavior in creation-of-the-universe schemes. The *International Journal of Theoretical Physics* is the last of Plenum's entries, and is, in some ways, the least tied to historically involved theoretical controversies. Not too surprisingly, it contains more particle papers than the other members in this assortment of Plenum titles, although most particles are discussed in a relativistic context.

The Plenum trio is challenged by two journals, one with a long historical connection and the other very recent. *Progress of Theoretical Physics* is a Japanese title (in English) that initially gained fame largely through the works of Hideki Yukawa and other particle theorists back in the 1930s. By contrast, *Classical and Quantum Gravity* is a British Institute of Physics title that has made its name only in the last five years, and has little direct discussion of particles, stressing fields instead.

Library selection in this overall group is probably best made on topical emphasis and the likelihood of your customer selecting a title as an outlet. Impact factors may be less important in this area than these two concerns. If your clientele is very heavily field oriented, *Classical and Quantum Gravity* and *GRG Journal* are first choices. If particle work albeit with a heavy relativistic emphasis is the case, taking the *International Journal* followed by *Progress* is a better strategy, although admittedly few non-Japanese authors publish in *Progress*. *Foundations* is a good compromise in small departments, particularly in collections that also serve a History of Science community.

### *Journals of Plasma Physics and Nuclear Fusion*

Four sources contribute to plasma physics literature today:

1. The straightforward physics and chemical engineering communities contribute from the tradition of an older, more encompassing field called fluid mechanics, where the physicists stressed gases as fluids and the chemical engineers defined fluids primarily as liquids. Both groups needed very involved mathematics to cope even fifty years ago. Today both groups see plasmas as specially ionized, generally superhot fluids with many special-case behaviors, requiring — it hardly seems possible — even more intensely involved mathematics, and increasing amounts of computer modeling in order to predict their flow.

2. The astrophysical community is involved because of its interest in stellar make-up and star life-cycles.

3. The electrical power community works with the hope of fusion reactors for electrical generation.

4. Finally, the materials science community deals with plasmas in vacuums for the creation of special surfaces as those plasmas are made to adhere to targets as thin films.

The journals in Figures 9 and 10 tend to represent differing mixtures of each contributing community.

Today's academic physics and chemical engineering departments approach to plasma science is best represented by two titles that had a parent journal devoted to all manner of fluid mechanics. These recommended offspring titles are the *Physics of Fluids B - Plasma Physics* from the American Institute of Physics, and the *Journal of Plasma Physics* from Cambridge University Press. The ancestor of

FIGURE 9

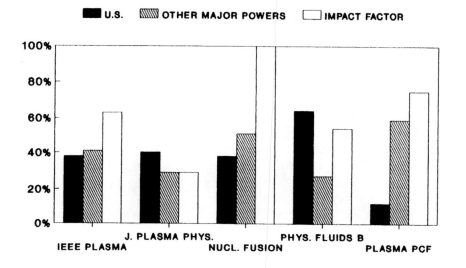

**Contributors and Relative Impact
Leading Plasma Physics Journals**

FIGURE 10

## Comparison of Number of Papers, Costs, and U.S. Market Penetration

the first journal continues today as *Physics of Fluids A - Fluid Dynamics*, while the parent of the second continues as the *Journal of Fluid Mechanics*. To a substantial degree the offspring journals derived from these parental lines also best represent the astrophysical community as well, although a huge number of papers in astrophysical journals are basically about plasmas in action in deep space.

The fusion-for-power-production community is best represented by the International Atomic Energy Agency's journal, *Nuclear Fusion* and by Pergamon's *Plasma Physics and Controlled Fusion*. This is the sexiest area in terms of impact factor, and both journals should be available in any strongly committed plasma physics lab or tokamak facility.

The *IEEE Transactions on Plasma Science* probably has the largest proportion of special materials processing papers in this group, yet also contains papers dealing with the electronic instrumental support of plasma fusion research facilities. This is very much like the way the *IEEE Transactions on Nuclear Science* supports accelerators. Another journal (not portrayed in these figures) and even more clearly committed to materials handling, but of some interest to other plasma workers, is Plenum's *Plasma Chemistry and Plasma Physics*.

If only one journal in this group could be supported, *Physics of Fluids B* is the best choice. The impact factors that appear weak in this figure will probably improve as plasma specialists realize more fully that they have now a society-sponsored subspecialty journal specially targeted to their needs.

### Journals of Astrophysics for Work with Particles, Fields, and Plasmas

Selection in astrophysics is very straightforward. Three journals absolutely dominate the field and concentrate the best papers between their covers. (See Figures 11 and 12.) Not only do these journals cover plasmas, they deal with particle detection (a historically significant number of particles have been first found via radiation from space), space-geometry and other field theory concerns. While small physics departments without a resident astrophysicist will not need all of them, most should have the University of Chicago - American Astrophysical Union title: *Astrophysical Journal*. *Astrophysical Journal* has a companion *Letters* section for short papers which is included in every subscription as well as a *Supplement* section for longer papers, available at extra cost.

Choosing between the *Monthly Notices of the Royal Astronomical Society* from Blackwell or *Astronomy and Astrophysics* from Springer on behalf of a broad consortium of Continental societies is largely a matter of scientific ethnicity and costs. The British title has a better price, but because some of its Commonwealth authors are from Third World countries, there are somewhat fewer papers from the more modern facilities. (As a rule, nowadays, astronomical observatories have become supported by consortial arrangements among several universities in cooperation with the financially better-off governments.) *Astronomy and Astrophysics* is quite expensive, but represents the best of Continental research. It is a highly commendable consolidation by an enlightened publisher of a number of formerly separate journals. If ever there was a journal which responded positively to librarians who complained about journal proliferation, this is it. It should be the second choice of most collections outside the Commonwealth.

FIGURE 11

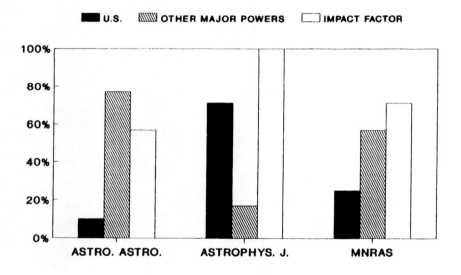

**Contributors and Relative Impact
Leading Astrophysics Journals**

FIGURE 12

# Comparison of Number of Papers, Costs, and U.S. Market Penetration

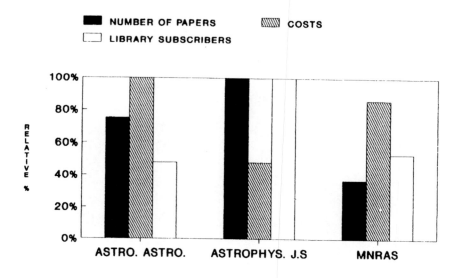

Chapter 6

# Journals for Contemporary Applied Physics

## BACKGROUND

### Who Defines Applied Physics?

Applied physics is a field whose topics and practitioners change very significantly over time. This means that there are fewer enduring themes and individually towering figures dominating the field much as Einstein dominated an academic concern like Grand Unification Theory. The agenda for applied physics journals can be reset somewhat unpredictably and by a wide variety of people. Today, the people who tend to control the agenda represent four constituencies. Each group has a slightly different angle on what constitute suitable topics for a journal in applied physics. Together the groups are arrayed in a spectrum of opinion ranging from those favoring works on fairly fundamental concepts to those favoring analysis of applied physics as seen in given types of industries and product lines. The spectrum includes:

1. *University physicists who work on "somewhat" applied problems, frequently with grants from the government or nonprofit foundations*. While certain fields like solid-state physics, optics, or acoustics have purely academic aspects, most of the research conducted within these fields, even at universities, has some potential for eventual application. This applied physics group understandably tends to favor lengthier, more scholarly papers with broad applicability as opposed to brief technical notes specific to one industry or type of product. These academic scientists still tend to be preoccupied with publications and grants as opposed to patents and con-

tracts. Moreover, these university professors of applied physics tend to gain tenure and promotions for such academic tasks as serving on editorial boards and acting as referees. Since their involvement is so strongly encouraged by their superiors, and publishers recognize that dedicated voluntary editorial help is hard to find, control over applied physics journals by the university contingent can be quite strong.

2. *Engineering professors at institutes of technology are also influential in applied physics journals.* They frequently consult for corporate or military clients. This can be more lucrative than not-for-profit research, but is in many ways more restrictive in terms of time and freedom to disclose findings. The engineering professor's entry into applied physics tends to be at the "prototype" level. He or she often develops early working models of more academic applied physics concepts. This group favors more papers on devices than do purely academic physicists. While involvement with journal publications and some editorial experience remain important for this segment of applied physics agenda-setters, contracts and patents have at least an equal status in their career advancement. Therefore, control of engineering school professors over applied physics journals tends to be more moderate than that of nonengineering faculty.

3. *There are straightforward, salaried technical managers.* They are most often corporate engineers who have day-to-day charge of a product or a process at a plant. They may follow the agenda of applied physics journals closely, but rarely determine that agenda directly. They seek the most practical papers of all, and favor comparisons of current technologies and materials. While these technical people rarely conduct sustained personal efforts at publishable research, they still feel a need to find ideas in these journals for making incremental improvements in their existing products and procedures. This group does spin off the occasional innovator with a start-up company. The history of a number of late twentieth century high-tech industries is based on enterprising individuals who were willing to make a change in a product, even when their employers thought it unnecessary or too expensive. When these salaried scientists-turned-entrepreneurs succeed, it creates a new industry that may change the applied physics agenda. Related to this

group of technical managers are applied physicists in hospitals and in materials testing facilities. Like their counterparts in manufacturing, they are not routinely expected to change the direction of the hospital or testing company but are employed for their technical knowledge of dangerous materials, powerful beams or complicated probes. Like their production manager counterparts, however, the occasional significant improvements they engender have launched a number of new companies that do have influence in the applied physics community.

4. *Finally, there is an exceptionally powerful contingent within the community of applied physics editors and contributors that represents a hybrid of all three previous scientific types: the long range major corporation "R and D" scientist.* The typical representative is something of a visionary who works on vaguely applied problems at a campus-like research center, but who is nonetheless a direct employee of the large company that sponsors such a center. Virtually every large American and Japanese firm, and many Continental European firms, have created research and development centers. There, both permanent staff and visiting production engineers interact at levels from basic concepts to finished products. These are generally the best-paid and most prestigious jobs within applied physics. The greater their achievement, the greater the research freedom and titles they are awarded. IBM's Watson Fellows, named after that firm's founder, Thomas R. Watson, are a prime example. Corporate managements have learned to lure some of the best academic talent to such centers by generous support of professional conference attendance and bonuses for journal article publications. Given this combination of factors, one can readily see how the agenda of many applied physics journals could be controlled by this segment. They have tended to align themselves with academics in favoring more long-range and fundamental papers.

## The General Political and Economic Circumstances that Alter the Agenda of Applied Physics Journals

The agenda of applied physics and its journals can be reset most directly by breakthroughs made while looking for a specific solution to a specific problem, or indirectly by applications found almost serendipitously for the earlier basic science work of academic scientists. Time lags between some basic discovery and its application are common, but these lags can be diminished by external forces. External forces such as wartime or economic competition often play a role in determining what constitutes acceptable applied physics. Perceived threats to a strong defense or to a healthy trade balance influence the funding of academic physics departments and private "think tanks." Competition among funding proposals serves as a general accelerant of the discovery of new applications for basic discoveries as applied physicists pore over older literature looking for ideas that will make their bids more attractive.

Until very recently the military dominated the agenda of the best applied physicists in both the U.S. and Russia. More than one hundred billion dollars a year annually supported the equipment and personnel of the research facilities of the defense departments of both countries—and this conservative figure excludes the actual management of weapons production and their deployment, a figure several times larger and including at least as many engineers in this second group as the first group included basic scientists. Two factors have forced a reevaluation of these expenditures. First, the decrease in Cold War tensions has made much of the military spending seem unneeded. Second, certain nations without an extensive military research commitment, but with considerable applied physics talent—most notably Japan—have gained tremendously in the applied physics of consumer products. This has so radically changed the economic balance of power that both the U.S. and Russia have been forced to alter their applied physics agenda so as to catch up.

### How Shifts in Trends Affect the Contents of Applied Physics Journals and the Assortment of "Must Read" Titles

Within applied physics, a topic gains value if the field is "hot" or in special demand. But, ironically, the journals of applied physics are as full of papers on the day before a "hot" breakthrough or new application as they are on the day after. Where do the journals get the room to publish the "new" hot papers? Where do the "cooled-down" papers find a home? While there is certainly some expansion of applied physics journals to accommodate new "hot" topics, the cooled down papers chill more slowly than might be expected.

The competing segments of the applied physics community rarely let them get frozen out of all kinds of publications, even if they no longer can hold their place in the most prestigious applied physics journals:

- The academics rarely let a program die entirely. They tend to send their chilled papers to the pure science outlets appropriate to the history of the field, as an example of how applications of the pure work are now possible.
- The engineering professor with a chilled manuscript can still work on problems of scaling up prototype production. He or she can submit the work to engineering journals that deal with aspects of general production engineering or industrial process design.
- The technical managers in given industries with chilled ideas still have more mundane trade journals specifically devoted to those industries. They will follow their topics there.
- The corporate "R and D" segment rarely has chilled topics. They are the ones who make the "hot" breakthroughs that chill all the other papers in the first place!

Consequently, shifting trends in applied physics actually sustain an ironic multiplication of journals, when research fashionability might be expected to reduce their numbers.

The following sections trace which topics or themes from the more academic and long-range corporate physics are finding employment and inclusion in applied physics today. Journals that are more fundamental to a wide range of applied physics collections will be stressed to a greater degree than those titles found only in engineering schools or within the manufacturing facilities of specific industries. In particular, the historical introduction that follows will help us trace how the contemporary situation came to be.

### The Applied Physics of the 1940s:
### Metals and Manufacturing

In the 1940s, applied physics stressed a number of concerns that today we regard as truly routine. Papers included the mixing of metals to form alloys, the use of X-rays to probe those alloys, and the testing of lubricants to reduce wear between metal parts. The vibration of metals and the handling of stress in metal joints were paramount. Aerodynamics, especially reducing the drag and turbulence of solids traveling in air, was moving slowly from a wind-tunnel science towards a combination of wind-tunnel and calculation. Energy concerns included the transmission of electrical power over ever longer distances. The development of frequency and power modulators for radio transmission and reception was extremely important. In the 1940s, this meant using vacuum tubes, a technology called thermionics, because the extreme heat of the filaments inside the tubes caused the production of electron flow. Interestingly the heat-based vacuum tube was heralded as far superior to the crystal-based radio technology of the prewar era.

Examples of other exotic topics included radar, the effects of radioactivity on metals and the means of tailoring synthetic rubbers for specific needs. World War II provided tremendous motivation and consensus concerning the importance and interdependence of all these topics. By the war's end, however, there was a reorganization of the agenda of applied physics, and the integration of topics fell apart.

Most metal physics had become metallurgy, pretty much a free-standing specialty out of the mainstream of applied physics until its reintegration in the 1980s through an alliance with materials science. Chemists and chemical engineers had taken over synthetic rubbers, and this area left the mainstream of applied physics. The field of electronic engineering, emphasizing communications technology, was well on its way to independence from its mother field, electrical engineering. The power generation portion of electrical engineering left the applied physics agenda, while the communications segment stayed and grew. The topics of vibration and mechanical wear and tear, which had been taken on by academic physicists as part of the war effort, returned to the more or less exclusive domain of mechanical engineering. Aerodynamics once again became a subspecialty of fluid mechanics and mechanical engineering. Nuclear engineering slowly became a civil engineering specialty incorporating a good deal of electrical engineering and fluid mechanics. This seeming narrowing of what constituted trendy applied physics opened the way for its domination by electronics.

## The Applied Physics of the 1950s: The Rise of Electronics and the Hunt for Semiconducting Materials to Aid Electronic Miniaturization

The 1950s saw a tremendous improvement in the ability of scientists to measure and calculate using electronic instruments. Probes of the fine structure of various new materials were improved using the electron microscope, an instrument developed twenty years earlier by academics. Ironically enough, old-style vacuum tube instruments were used to develop the category of electronic materials that were eventually to make vacuum tubes obsolete in many circumstances. The "old" thermionic vacuum tubes that "lived" by extreme heat "died" by extreme heat. The filaments inside the tubes burned out — often when the air seals dried out. Up until the 1950s, the applied physics agenda was pretty direct: to find new materials for filaments that could stand the heat longer, and to develop tighter vacuum seals. In the 1950s, the approach became indirect: miniaturize the tubes. With less current required, the heat build-up would be less, and the circuits would last longer.

But something unexpected happened on the way to smaller tubes. The hollow tubes were replaced by solid transistors. A previously overlooked solid element, germanium, sometimes mixed or layered with other elements, cut down the size of components in circuits at least tenfold, and extended the lifetime of the circuit board at least that much. Instead of fiberboard panels measuring a few square feet, with bulky wires connecting sockets for salt-shaker-sized tubes, one found plastic resin or ceramic boards a few inches square with fine electrical wires attached to small germanium transistors.

This initial solid-state era of miniaturization was due primarily to a trio of researchers who went on to win the 1956 Nobel Prize. These were John Bardeen, Walter Brattain, and William Shockley, all initially employees of Bell Labs. While each has gone on to further notoriety in science, their paths have been quite divergent. Bardeen took a senior professorship at the University of Illinois and went on to win a second Nobel Prize. Brattain devoted many years to a quieter level of research, and made an exceptional commitment to undergraduate teaching by returning as a professor to his alma mater, Whitman College, a very small school in the Pacific Northwest. Shockley worked for a number of corporations before signing on with Stanford. There his career in physics was overshadowed by his espousal of racial theories of intelligence, and by his highly publicized contributions at a rather advanced age to a sperm bank for the procreation of genius babies! It is not clear whether or not his deposit has gained much interest.

The solid-state miniaturization boom greatly influenced the growing computer industry, particularly the entrepreneurial start-up segment of applied physics. An Wang (An is a male gender name in this case) was a Chinese immigrant who had worked his way through Harvard, and set up an electronics shop in his garage. He suggested that new components with electrically switchable magnetic polarities might be incorporated into these newly fashionable circuit boards. Given patterns of micromagnets switched "on" or "off" could stand for text or numerals in a much more compact form of computer memory than was then currently used. The idea led to Wang Computers, and this magnetic core memory technology was to serve the 1960s well. It also led to increased emphasis on magnetic materials within materials science. Ken Olsen, an MIT graduate working out of an abandoned textile factory in the mill town of Maynard, Massachusetts, was one of the first scientists to see that this trend towards miniaturization meant that computers could be built and sold to smaller firms and more modest colleges. Before Olsen, the typical computer was room size, and was generally sold to a large research university or to a Fortune 500 business. Olsen's minicomputers were the size of refrigerators and were affordable for just about every college and medium-size business in the country. They still sell, and continue to shrink ever smaller, under the brand name of Digital.

During the 1960s the communications aspect of electronic technology tended to leave applied physics while the computer component aspect increased dramatically. Component physics focused increasingly on materials like germanium that had semiconducting properties — that is, they conducted electricity somewhat less well than pure metals, and somewhat better than insulators like polymers and ceramics. This gave them great potential as modulators and switches of electronic signals.

Momentarily out of research fashion, the ceramics industry joined metalworking and the plastics field as virtually independent engineering disciplines with an exception important to applied physics literature. Unlike the situations of metallurgy and polymers, there were fairly few graduate programs in ceramic sciences, and even fewer journals in the field. Indeed, ceramics formed a natural alliance with inorganic chemistry that was later to lead to the reemergence of ceramics in the applied physics mainstream.

## The Applied Physics of the 1960s: Lasers and Ultrasonics Join Electronics as Hot Topics

While progress in computerization and related experimentation with semiconducting materials would continue apace, the 1960s saw two fairly unexpected new developments. The first and most dramatic event was the rise of the practical laser out of earlier, more academic work on masers, a microwave-based antecedent.

The earliest academic work on masers was Russian and initially not well-understood in the West. The underlying idea was to develop the ability to focus rays of exceptional uniformity and power over distances with little spreading out of the beam: a theatrical spotlight effect as opposed to a parking lot floodlight. Nicolas Basov and Alexander Prokhorov were employed in Soviet research centers and had made a number of convincing arguments that the vibrational energies of molecules — long noticed by basic scientists as being extremely constant and dependable — could be harnessed to make these powerful beams. The harnessing had to be in a manner that was self-reinforcing: one excited molecule should excite another to vibrate at exactly the same wavelength. Of the many wavelengths of the electromagnetic spectrum at which to attempt this synchronized excitation, Prokhorov and Basov suggested microwaves based on theoretical calculations and some early experiences in radar, which uses microwaves.

A practical maser awaited Western development, in part because the electronic components necessary for energizing and synchronizing the vibrations were better developed in the burgeoning and competitive American electronics industry than in well-funded but less innovative labs of the Soviet academies and military centers. Ted Maiman of the Hughes Corporation and Nicolaas Bloembergen of Harvard University made the first working American masers in the late 1950s, using excitable gases to generate the beams. By the 1960s, the resonating, coherent wavelengths shifted from the invisible microwave to the visible light ranges of most of today's lasers. The most successful of several competing claimants of laser discoveries were Charles Townes, who worked for Columbia University at the time of the discovery, and Arthur Schawlow, who went from Columbia to Bell Labs just before the discovery. With time, lasers were engineered to have solid-state cores with a number of advantages in size and stability. Soon, liquid cores served as a source for excitable material. Newer cores allowed a choice of wavelengths at given settings on the dial, so that the most effective frequencies for different applications could be serviced by a smaller assortment of lasers.

The early expectations of the laser were that its uses would be primarily in military weapons (death rays!) and for purely scientific probes. The surprise was that their uses would quickly become much more widespread and commonplace. Indeed, to a substantial degree yesterday's maser would become today's microwave oven, and there are more lasers today in photocopy machines and bar-coded supermarkets than in all the armies of the world. By the way, the regularity of the vibrations of atoms and molecules on which Prokhorov and Basov speculated has been incorporated into the finest timepieces of today. The scientific clocks at places like the National Institute of Standards and Technology depend on the internal beating of the cesium atom in maser-laser fashion.

The burgeoning health care sector would prove to be another excellent stimulus to applied physics research. While laser surgery would become common in the 1980s, the 1960s saw the first significant uses of another even older wave technology: acoustics, the science of sound and vibration. Ironically, the acoustics that made the breakthrough was not in relation to hearing deficiencies, it was in relation to less dangerous ways of probing bodily structures and function.

As far back as the turn of the last century, English researchers such as James Strutt, later and better known as Lord Raleigh, had noticed that very high-pitched sound could cause a flame to flicker rhythmically. (Certain opera stars can register high and strong notes that can cause glass to vibrate to the shattering point.) He tied in this observation of an invisible and often inaudible vibrational energy effect with long-standing observations from nature. He noted the abilities of cats and owls to hear the high-pitched squeals of mice, and of bats to not only hear such sounds but to generate them and use their return echo to locate prey. Raleigh's flame detector for ultrasonic waves was impractical outside the lab, and he was not able to make many successful ultrasound generators, and he could only crudely record some naturally occurring ultrasonic phenomena.

This applied physics achievement would be stimulated by wartime. Ultrasound (sonar) to detect submarines was developed by Paul Langevin, a French scientist who had already achieved fame as both a leader in early relativity and in promoting radical French socialism. (Einstein said that should he himself have not proposed relativity, Langevin would have. Einstein also lauded Langevin's later heroic resistance to the Nazis: Langevin's son-in-law was executed and his daughter survived Auschwitz.) Langevin was one of several scientists who hit upon the idea that electrical charges could make specially tailored crystals vibrate at highly predictable speeds, and that the vibrations could be stepped up by increasing the current. (Quartz watches work very much on this principle.) Langevin was the key scientist in scaling up the size and power of both the ultrasound generator and its echo receiver.

The early military direction of applied ultrasonic physics was not favorable to its later medical applications. The pitch and intensity of ultrasound always seemed to be directed to higher energies and greater power: by the 1950s the waves could penetrate railway beams and detect cracks in the ironwork. Indeed, focused ultrasonic beams were developed that deliberately caused a kind of welding.

The detection of large scale ultrasonic vibrations became a basis of oil well exploration and earthquake monitoring.

Fortunately the fine-tuning and electronic miniaturization schemes of the 1960s turned down the volume of the ultrasound, so to speak. A partnership among electronics engineers, physicists employed in hospitals to handle X-ray and radioactive isotope equipment, and physicians specializing in radiology helped the progress of ultrasound along considerably. Among the earliest biomedical applications of ultrasound was the separation of body parts and chunks of tissue into individual cells by a kind of ultrasonic blenderizing: the vibrations basically shook loose the attachments among cells. Turning down the volume further, physicians found that nondisruptive ultrasonic waves could take "pictures" of softer tissues like muscle, that X-rays often shot clean through without revealing much detail. The most dramatic application of medical ultrasound has been in obstetrics, where babies can now be safely monitored throughout their development in the womb. Nonetheless, other important applications have included the measuring of flow of body fluids through valves in the heart, gall bladder, or kidneys.

### The Applied Physics of the 1970s: Chip Wars, Fiber Optics and Image Storage

The quest for ever smaller, ever cooler, ever more densely packed circuits pushed the development of integrated circuits. The old palm-sized board of transistors was replaced by a finger-nail sized chip. The new set up required a new matrix for the circuitry. The old plastic or ceramic boards with germanium components were replaced by the silicon chip. Ultrapure silicon wafers manufactured in ultraclean facilities became the main product of areas of California, Massachusetts, Texas, and Japan. Ironically enough, this prompted a revival of crystal growth technology that harkened back to the original radio technology. Wafers of crystalline material had to be cast and cut and engraved with anchoring sites for the circuits and components that would make them run as units of electronic memory and logic. Initially this was done with mechanical inscribing with wiring emplaced by skilled workers using visual light microscopes to guide them. Later on, a kind of photochemical engraving would deposit ultrathin alternating layers of metal and insulation for circuitry using X-ray lasers piercing through pattern grates to define the layout. It would soon take an electron microscope to visualize the chip as an intricate city of tiny electronic houses and streets, with virtually all utility attachments to the chip city now done robotically.

Optics continued to make advances that were critical to the agenda of applied physics. Two advances in particular fueled the industry. The first was fiber optics.

Fiber optics describes the technology of light waves traveling within strands of plastic or glass tubing with no escape of that light tube to the outside world until its destination was reached. While the basic principle was discovered as far back as 1955 by an Indian physicist, Narinder Kapany, the widespread application would await two developments of the 1970s. First, very long, strong, and thin strands of glass fibers had to be economically produced, largely a triumph of America's communications industry corporate research centers. Second, lasers and other high intensity light sources had to improve, largely through lower cost mass production and the ability to become pulsing units, triumphs of scores of small start-up firms throughout the U.S.

The second major advance in optics was the widespread incorporation of magnetic tape, a medium hitherto favored for recording music or computer data, as a medium for storing visual images. Videotape was a rudimentary technology started in the early sixties. The core idea of videotape was to make films of breaking news events more quickly available to television stations without the photochemical processing of conventional film. The original videotaping equipment was only slightly smaller than a conventional television camera and could weigh nearly one hundred pounds. The playback equipment was the size of a kitchen table. Nonetheless, videotaping was much faster, if not initially more sharply focused, than the "wet" film technology. Beginning in the 1970s, the use of microcircuitry and micromachinery required only battery power, enabling the recorders to become a great deal smaller. Moreover, the technology of magnetic tape greatly improved, allowing much greater density of visual images in each "frame" and leading to both a sharper image and much shorter and narrower tapes. Eventually these tapes would fit into convenient cassettes, much as today's music and computer-backup tapes do.

Stimulated by this pragmatic development — and the rise of faster and denser computer chips to encode digitized images — optics expanded its scope to deal with the digital storing, retrieval, and enhancement of visual information. It took on the added specialties of image science and pattern recognition in the 1970s. This expanded definition of the field came just in time to help improve acoustics and other areas of medical physics.

The progress in ultrasonics during this decade focused on ever sharper refinements of the visual image generated through computer conversion of ultrasonic feedback to video. Once again there was more use of computer chips and magnetic disk memories to capture, store, manipulate, and reproduce images for later study. Moreover, with better targeting information and by once again fine-tuning and learning to pulse the power of ultrasound, some deliberately disruptive ultrasonic shock waves could be aimed for the beneficial destruction of stones in the gall bladder and urinary tract. A significant number of highly invasive surgeries could now be avoided.

The biggest news in health physics, however, was the resurgence of X-ray technology, particularly the "CAT" scan. Computerized Axial Tomography is a means by which X-rays are taken at various angles through the body, with the composite figure being averaged to give a near three-dimensional construct of some body region. (The two-dimensional print that we see in medical journals resembles nothing so much as a body sawed in half!)

This medical imaging triumph resulted in a much deserved but peculiar Nobel Prize in Medicine for Allan Cormack, currently of Tufts University in Boston, and Godfrey Hounsfield of England. Neither researcher was a medical doctor; indeed, neither had an earned PhD! Cormack was essentially an applied physicist with a master's degree who emigrated to the U.S. from South Africa. He contributed the key equations on reconstructing 3-D images from films shot at several angles. Cormack's most significant work up till then was in instruments for nuclear physics, not in the hospital physics that was to win him the prize. Hounsfield was very much a working electronic device maker at a highly reputable but rather modest-sized scientific instrument company he had founded. He personally designed and constructed the prototype multi-angle X-ray machine. A key to the design and construction was the use of an array of small detectors with computer chips, instead of actual X-ray films, to capture and encode the multiple images. With high speed processing using Cormack's equations, images that could later be even further manipulated or enhanced for surprisingly revealing details were produced.

### *The Applied Physics of the 1980s and 1990s: New Materials, Great Integration of Contributing Specialties, and the Phenomenon of Superconductivity*

The solid-state electronics revolution reached the consumer full force in the early 1980s. Ever more densely packed computer chips and hard disks, ever more reliable mechanical drives, cheaper magnetic diskettes, and laser and fiber-optic printers put high-quality, inexpensive desktop terminals in tens of millions of homes and offices. Few industries show the interdependence of areas of applied physics as the personal computer and workstation business, and few were better examples of the rise of influence of the entrepreneurial "start-up" segment within applied physics. Ultimately, however, the major long-term applied physics advances of this decade were dependent on the nexus of the major corporate research center and university professors of applied physics.

The best known advance was the rebirth of interest in superconductivity. The initial discovery of the superconductor effect (the virtual disappearance of resistance to the flow of electricity in supercold materials) was due to a Dutch researcher, Heike Kamerlingh Onnes, in 1911, clearly a case of a long time between basic discovery and rush to application! Kamerlingh Onnes was a student of the famed German scientists Bunsen and Kirchhoff, and was the most successful scientist of the supercold since Lord Kelvin. (The last three scientists mentioned are discussed extensively in the chapter on physical chemistry.) Kamerlingh Onnes was also one of the most successful famine relief workers of the post World War I era, and, like Kelvin, he maintained a strong interest in the practical refrigerator. Kamerlingh Onnes was able to liquefy and freeze gases where none thought it possible.

The problem with his superconducting effect was that it was impractical. Maintaining supercold temperatures was expensive. Each degree closer to absolute zero was hard won and more difficult to sustain. Early electrical devices within this environment had problems with their moving metal parts breaking from brittleness in the cold. The advantage of no heat build up in electrical circuits — the Holy Grail of electronics technology — seemed possible but hardly worth the effort.

But technology had obviated some of these difficulties by the early 1980s. Computer chips, for example, have no moving parts. During this time Bednorz and Muller of IBM's Swiss long-range research center also got around the requirement for the most extreme cold for the zero-resistance effect. They developed new materials that showed superconducting behavior at a dozen degrees higher than Kamerlingh Onnes had ever seen. A dozen degrees meant more than it first appeared. There was already a considerable savings of effort for many working scientists, and a Nobel Prize for Bednorz and Muller. Their success set off a competition with the rest of the world to find materials that would exhibit superconductivity at even warmer temperatures. Indeed, Paul Chu of the University of Houston became regarded as something of the master of superconducting materials cookery. The epithet "cookery" is used aptly, for although the 1972 Nobel Prize-winning theory of Bardeen, Cooper and Schrieffer gave a workable explanation of what occurs in superconducting materials, there are, as yet, relatively few general formulas or structures that will guarantee a workable high-temperature superconducting unit.

The search for superconducting materials has forged a new link between solid-state physics and inorganic chemistry, since the most promising composite materials seem to contain some mix of copper, yttrium, barium, bismuth and oxygen — all elements within that older specialty's domain. Moreover, many promising superconducting materials turn out to have ceramic-like structures and require ceramics technology, making microceramics once again a hot area within applied physics.

The solid-state electronics industry is not betting all its cards on superconducting materials, however. Just as early silicon crystals were displaced by tubular germanium transistors, which in turn were replaced by essentially flat silicon-based chips, so now a new category is being actively explored: gallium-arsenide doped materials. While this new material is far less exotic than superconducting mixes, and the gallium-treated materials will likely be embedded in a familiar matrix of silicon, gallium's payoff industrially is much closer than that of superconductivity. In practical terms, gallium technology will enable the construction of even denser chips — gallium-doped transistors as incorporated into microchips are a hundred times more efficient than existing etched-in chip transistors — with even greater use of Josephson junctions. Interestingly, Josephson junctions, named after British physicist Brian Josephson, were explored as far back as the 1960s, and allow for the design of ultrafast electrical switching on chips with trivial heat buildup, precisely the need of high speed computing. Most curiously, after achieving enormous fame and a Nobel Prize for work done at the age of twenty-two, Josephson became seriously involved with Eastern mysticism, particularly the beliefs of the Maharishi Mahesh Yogi.

His fellow applied physics laureates of that year have continued on in more conventional careers. Leo Esaki, a career designer of ultrasmall components, left Sony for IBM, where he remains today. Ivar Giaever, who, like Einstein, began his career as a patent examiner, is now with General Electric. Giaever personifies the continuing link between those who work with electron microscopy and those who study the very thin materials with special properties that are dominant in today's electronics. Indeed, two new technologies, electron force microscopy, and scanning tunnelling microscopy are seeing applications in this area.

The 1980s were particularly good to the laser segment of optics in the implementation of laser-optical storage. In this method of storing images or text, an element of a picture or printed character is etched onto a plastic disc. A laser reads the hills and valleys on the disc and reproduces the image. This is conceptually very similar to CD-ROM and has advantages in picture portrayal if not in text storage just now. Its costs are not as favorable and equipment distribution at this time is not as widespread as videotape or CD-ROM, but its potential for greater storage remains very high.

The 1980s saw another advance in improved medical imaging:

nuclear magnetic resonance. Much like CAT scans, NMR scans slice through a patient and can, with integration of computer graphics, give something like a three-dimensional image. The major advantages of NMR are that it involves no penetrating radiation to speak of and that it can monitor something of the physiological state of an organ. Rather than sending an energetic beam like an X-ray into the body to look for information from collisions with the beam, NMR essentially listens to the body's own atom-level broadcast. It does this by picking up the shifts in signals sent out by the nuclei of atoms within the body that occur in response to the strong magnetic field in which the patient is placed. Patients undergoing a medical NMR scan are placed in what seems very much like a tube. It is unlike CAT scans, which also place the patient in a restrictive tube, in that there is no X-ray camera revolving around the patient. Instead, the strong magnetic field envelopes the patient so as to create the conditions for this pick up of the shifts in nuclear signals. With NMR not only are the organs pictured but some information about the fluids in which they function can be obtained. Ironically, the fluids information is often well understood, despite the relative newness of medical NMR. NMR has largely been a tool used by chemists for over forty years to identify liquid compounds involving hydrogen, carbon, or phosphorous. (Most medical NMR works by monitoring the signals from the hydrogen atoms in the water within the internal organs.) Nonetheless, there is some historic justice in the return of NMR to the physics community, since the phenomenon was first discovered in the 1940s by Felix Bloch, a Jewish refugee and nuclear physicist who settled at Stanford, and by Edward Purcell, a nuclear physicist who worked largely at MIT and Harvard.

### The Component Disciplines of Today's Applied Physics: Solid-State Physics

The quasi-academic field that has dominated physics for the last thirty years, solid-state physics, will continue its leadership role for the foreseeable future. Solid-state physics is the study of the properties — particularly the electronic and magnetic properties — of matter in its condensed "solid" state. While the oldest solid-state physics materials were natural crystals and metals, the most studied materials today are synthesized in the laboratory, usually from molten materials. Most of the materials are composed solely of noncarbon materials, and are therefore called "inorganic" solids. But research into some carbon-based solids that conduct electricity, an area typically called "synthetic metals," is growing. The key idea of solid-state physics is that the relative orderliness of crystalline materials allows for the electrons that make up electricity to flow along — or as it is commonly termed, to "tunnel" — at very high speeds through the lattice-work with less impedance and heat build up than within more disorderly "noncrystalline" solids or fluids.

## The Component Disciplines of Contemporary Applied Physics: Material Science and Surface Science

Materials science is a field that is older than solid-state physics, and yet solid-state physics is certainly the best customer for the findings of materials science. Materials science is a discipline that has its historic roots in engineering, particularly in such practical questions as what is the strength of a given type of concrete, what is the heat resistance of a given wire, what substances are good at waterproofing a fabric. The difference between materials science and one of its parent fields, analytical chemistry, is that materials scientists not only analyze the material, they seek to tailor its properties through treatments or through alternative methods of formulating it. Materials scientists have influenced solid-state physics by two seemingly opposite methodologies. First, they attempt to produce ultrapure, contaminant-free materials and study their properties. Alternatively, they have become experts at exploiting the use of induced "defects" and controlled amounts of "impurities" in semiconducting materials to get desired effects. Most advanced materials today involve layers or composites, and can be produced in several versions for differing properties. Yet the ultrapure segment is making a comeback with products like ultrahard and ultrathin diamond films.

The older, established specialty that has achieved the greatest revival through its alliance with materials science is metallurgy. Scientists originally trained in metallurgy alone today view themselves as more than steelmakers in much the same way that enlightened transportation and shipping companies see themselves as more than truckdrivers, railroad workers, or dockworkers. They have an expanded agenda for their firms and their individual research efforts. They have developed both products and collaborations involving polymers and ceramics and semiconducting crystalline solids.

The new specialty within materials science that has grown most rapidly is surface science. The electronics industry in particular has made extensive use of thin solid wafers and has kept material scientists busy developing ultrathin and ultrapure coatings. Chemistry, particularly colloid chemistry, has been a contributor to surface science, in part because the special properties of thin wet films exhibit much commonality with the surfaces of colloids in solution. (See the chapter on physical chemistry.) Moreover there are many topics in common between colloids and surface science such as corrosion, membranes, filtration, and electroplating. Indeed, within the United States, Nobel laureate and GE researcher Irving Langmuir was effectively the founder of both colloid chemistry and surface science.

Materials scientists are extremely eclectic in their use of tools and research methods. They have taken plasma physics and turned it into a means of making very pure and very thin materials within vacuum chambers. They have taken low energy nuclear physics to not only probe the surfaces of solids with beams, but to coat those surfaces with "sputtered" atoms shot out from other materials bombarded by even stronger beams. Surface scientists can routinely make layers only a few atoms thick.

Materials scientists do have clients other than in the electronics industry. Most of today's "stealth" technology, bullet-proof glass, biodegradable diapers, and microwaveable kitchenware, are due to the formulations of materials scientists within the metals, plastics, and ceramics communities.

In recent years, the biomedical community has become a significant client of materials science. Artificial limbs and implanted body parts constitute exceptionally demanding problems in materials science. The materials must not irritate the body, must be able to be shaped to the individual size of the patient, and must last at least as long as the patient's lifespan. Indeed, the biggest crisis of contemporary biomedical materials science is that some body parts, such as heart valves and artificial orthopedic joints, are wearing out for an unanticipated reason. The lifespans of the recipients have been so successfully lengthened that the "warranty" expectations at the original time of surgical installation have long been exceeded.

## The Component Disciplines of Contemporary Applied Physics: Optics and Acoustics

Optics today has many more roles than Isaac Newton ever dreamed. Optics continues in its traditional role of allowing visualization, and the resolution of today's microscopes is down to single atoms. (IBM recently photographed its corporate logo done in single atoms using scanning tunnelling microscopy). Fiber optic scopes that can be inserted into the body — endoscopes — allow doctors to see internal organs without massive incisions or more dangerous X-rays.

Optics now plays a role as a carrier of information. It uses the modulation of wave behavior to bear voice and data messages. In recent years, virtually all new installations of telecommunication cables in the industrialized West have been of optical fiber cable, rather than old copper wire. Very fast computer workstations now often feature photonic rather than electronic interconnections.

Optics has an ongoing role as a cutting tool. X-ray lasers have begun to substitute for etching instruments in cutting out the pathways of the ubiquitous computer chip. Optics also has a role as a robotic scanner or sensor. It may surprise the reader that many food processors optically scan foods like peanuts on a conveyer belt. Sensors kick out insects, pebbles, and off-color nuts caused by aflatoxin, a fungal disease.

Optics has an important role in national defense and economic competition. Remote sensing of the environment is done by long-range optical detectors mounted on satellites. Reflectance of sunlight from the plains of the Ukraine can tell us if we will be selling wheat to the Russians in the coming months. Ultrafast spectroscopy of exhaust contrails can tell us if an airliner is on the way or if we are facing a missile strike.

Optics can perform more than one role at once. Today, not only can a soft optical beam map contours of the eyeball, but it can power up in some cases to either fuse a detached retina through laser surgery or laser-guide the cutting tools at the computerized optician's workbench so as to make a corrective lens. Most large components in car manufacture are assembled and welded using laser guided robots. Opto-electronics, the coupling of optics with integrated circuitry, is leading to hybridized devices that are at once sensors and processors of external information, with some reflectance data effectively entering itself onto the chip, bypassing the typing in of numbers. Optical engineers are among the most important workers in robotics and artificial intelligence. Together, computer scientists and optical experts are sorting out just how much visual information a robot or computer needs to ''see'' in order to be able to decide on a course of action. In humans, this study is a part of a growing field called ''pattern recognition,'' and it plays an important role in designing the workstations of the future.

While acoustics has shown less spectacular results than optics, it remains a significant field within applied physics. Ultrasound remains by far the most seriously employed aspect of acoustics, particularly within medicine, but the multibillion dollar home musical entertainment industry still accounts for more dollars. Curiously, architectural acoustics and the remediation of deafness, traditional concerns of acoustics, have still not been reduced to simple formulas. There remains a blending of intuition and trial-and-error along with computer generation of models to describe what is going on. In fact, general journals of acoustics still frequently feature papers from both musical and audiological communities alongside those from applied physicists.

Acoustics has never shaken off its connection with underwater sound detection. Sonar is as important to submarine monitoring today as ever. Vibrational analysis—indeed, one forgets that audible sound and ultrasound are but two of many examples of vibrations— continues to occupy mechanical engineers. Machines not only generate vibrations, but those vibrations cause wear, mechanical failure, and unwanted noise which often requires damping. Of course other acoustics machines are used to deliberately generate vibrations and then monitor the response of the ground. Acoustic feedback can tell a structural engineer about the strength of a bridge or tell a geophysicist if there are discontinuities, like oil or water pockets, several layers beneath the surface of the earth.

Acoustics has become, nonetheless, a full partner in the microelectronics revolution. As television ads continually remind us, miniaturization has certainly come to the hearing aid. What is less realized is that many of today's autofocusing cameras use a kind of sonar to judge the distance to the subject of a photo and automatically adjust the focus, a blending of solid-state electronics, acoustics, and optics. Indeed, one of the most demanding areas of acoustics research as it relates to computing is the recognition and processing of speech, a field very much equivalent to the optical scientist's pursuit of pattern recognition. This blend of the contributing specialties of applied physics is characteristic of ongoing research in each of the fields, and is one of the persistent themes for applied physics overall in the 1990s.

## The Component Disciplines of Contemporary Applied Physics: Diagnostic Imagining, Beam Therapies, Managing Radioactivity

Perhaps the most adaptable community of applied physicists is within biology and medicine. The group began the century with the dangerous but fairly straightforward handling of X-ray machines and radioactive isotopes. It has, in the last twenty years, added particle accelerators for cancer therapy, ultrasound, lasers, fiber optic endoscopes, CAT scans, and magnetic resonance imagining. Not surprisingly, given the graying of the American population and its consequent need for more medical care, hospital physicists are likely to increase in numbers. While on the surface it may be argued that the new machinery and tests they supervise are very expensive and contribute to the rising cost of health care, it can be shown that there has been significant cost savings through the straightforward reduction of the number and risk of invasive procedures. Consider that fiber optics has obviated the need for much exploratory surgery, and reduced the size of incisions necessary for knee surgeries and gynecological interventions. Not only has ultrasonics enabled us to envision gallstones, but to pulse shockwaves to break them up in an increasing number of cases. The amount of radiation from isotopes and X-ray machines has been reduced by CAT scans, and eliminated in NMR. While some cancers are still being treated by contact exposure to radioactive materials, an increasing number of cancers may be treated by using particle accelerators that can be aimed precisely, and whose intensity and dosage can literally be programmed into the machine. Moreover, after the machine is shut off there is virtually no radioactive waste.

The health physics segment of applied physics — and there is some distinction between them and the hospital physicists — is also likely to see increased employment as concerns about environmental radioactivity mount. This is due not only to more old nuclear power plants being decommissioned, but also to more toxic wastes sites being uncovered, and the rediscovery of the hazards of radioactive gases like radon accumulating in the basements of many homes with certain stone foundations. (Health physicists had initially studied radon's dangers as consultants to mining engineers.) This health physics group has accumulated expertise in the proper handling of these materials and has become particularly inventive in devising detectors for radioactive leaks and in decontamination procedures.

Of course, not all radiation uses are bad, nor are all applications of radioactive materials unintentional. Radioactive materials and radiation-emitting machinery will continue to play an important role in many materials testing and manufacturing procedures. Health physics officers will continue to generate the safety literature that must accompany these efforts.

## SORTING OUT THE JOURNALS

### Journals Covering All Areas of Applied Physics

Much like the world of high technology, the world of general applied physics titles is becoming increasingly competitive. As Figures 1 and 2 indicate, the U.S. titles *Applied Physics Letters* and the *Journal of Applied Physics* (both from the American Institute of Physics on behalf of the American Physical Society) each have some leadership aspects, but are being challenged. *Applied Physics Letters* is the impact factor leader, and the *Journal of Applied Phys-*

FIGURE 1

# Contributors and Relative Impact
# Leading Applied Physics Journals

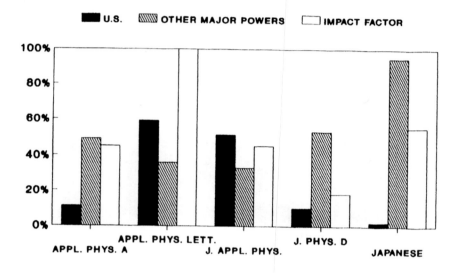

*ics* is the leader in total number of papers. But the impact factor of the *Japanese Journal of Applied Physics* (and its included *Letters* section) is quickly moving up, even as its total number of papers increases. This quality-with-quantity phenomenon is quite curious since that journal has announced that, for all practical purposes, it is foregoing refereeing (the standard quality control method) so as to keep the flow of papers undeterred. One would expect that this would result in a large number of lower quality papers that would be rarely cited, yielding low impact factors, but this has not turned out to be the case. The self-discipline in submitting only high quality papers on the part of the Japanese scientists involved — and virtually all the papers in this English-language journal are from Japan — is remarkable.

Yet another remarkable development is the surprisingly good

FIGURE 2

## Comparison of Number of Papers, Costs, and U.S. Market Penetration

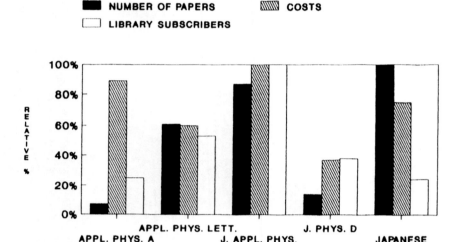

showing in subscription loyalty of American libraries to the British entry, the *Journal of Physics D - Applied Physics* from the Institute of Physics, relative to its German competitor, *Applied Physics A - Solids and Surfaces*, from Springer. Language cannot be the reason since the German title is virtually entirely in English. Price-per-paper and tradition may have more to do with it, since the rate of American participation as authors in these two foreign but English-language journals is about equal, and the British impact factor is markedly poorer.

While virtually all serious collections in applied physics should have every title analyzed in this section, the American titles are an absolute must in even the smallest collections. The Japanese title is a good next choice, and despite its unfavorable financial aspect, the German title is to be preferred over the British title. The long-term cost of losing intimate contact with German developments in applied physics will probably be greater for one's institution than the short term savings in sticking with the British subscription.

Review journals are scarce in general applied physics. It might be argued that by the time a comprehensive, evaluative review paper is written, the technology will have changed so much that the paper will have diminished worth. Nonetheless, it should be noted that the *Journal of Applied Physics* publishes a fair number of reviews within its regular issues, and, on a recurring basis, devotes special issues to symposia papers in "hot" areas such as magnetic materials.

### Journals of Solid-State Physics

As has been repeatedly emphasized in this chapter, solid-state physics, particularly through its connection to the electronics industry, is the dominant area of applied physics. Virtually no collection in applied physics could reasonably function without several titles in solid-state physics. Once again, the U.S. titles have some dominance in the international arena, but they are being challenged.

As Figures 3 and 4 indicate, the impact factor leader is the German title *Zeitschrift fuer Physik B - Condensed Matter*, an entry from Springer. Some critics would argue that this is a fluke owing to the thousands of citations to the Nobel Prize winning work of the Europeans Bednorz and Muller on high temperature superconductivity. They suggest that this impact factor may not represent any lasting superiority of the *Zeitschrift* over the American entry *Physical Review B - Condensed Matter*, and the German impact is certain to moderate.

But one might pause to reflect on why Bednorz and Muller chose this outlet in the first place. As employees of one of IBM's long-range research facilities in Europe, they certainly had ample resources, reputation, access, and familiarity with the American Institute of Physics/American Physical Society title, *Physical Review B*. They simply felt that the more broadly European outlet, the *Zeitschrift*, was at least as natural a choice for this European work as the perennial American leader. Given the growing sense of cooperation among individual European countries and the increasing sense of European researchers not as citizens of individual European states but as participants in the European community, it might not be too

FIGURE 3

## Contributors and Relative Impact
## Leading Solid-State Physics Journals

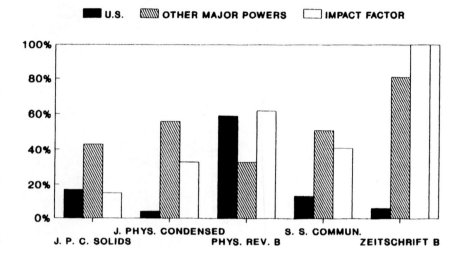

surprising to find that other top-notch European researchers also choose the *Zeitschrift* and sustain, to some degree, its healthy impact factor. Consequently, the *Zeitschrift* should be in all but the smallest solid-state collections.

Nonetheless, *Physical Review B* remains a clear choice for all physics collections of any size whatsoever. It marshals between its covers more of the developed world's solid-state papers than the next four competing journals combined. Even schools with an applied physics program but without a particular solid-state emphasis should take this title, because it reports so much that impinges on other areas of physics that it is essential to their study as well.

The next selection after the *Zeitschrift* and *Physical Review B* is also clear. *Solid-State Communications* from Pergamon has become the principal vehicle for the world's short, quickly published papers in this specialty. It has a fair number of American papers and, in

FIGURE 4

## Comparison of Number of Papers, Costs, and U.S. Market Penetration

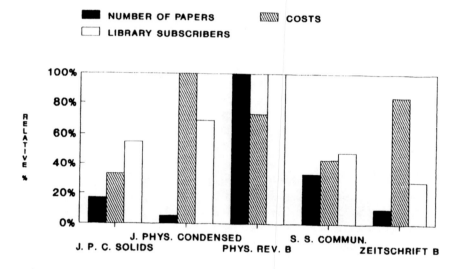

■ NUMBER OF PAPERS     ▨ COSTS
☐ LIBRARY SUBSCRIBERS

fact, among solid-state journals, publishes the second most American papers after *Physical Review B*. Its costs per paper are surprisingly competitive for a for-profit title, although the briefer nature of the papers skews this factor somewhat in the publisher's favor.

The final recommendations in this group are less clear, and are influenced by the subject emphasis at the host institution. If the resident solid-state physicists have close ties with solid-state chemists and crystallographers, then the *Journal of the Physics and Chemistry of Solids*, the "parent" of *Solid-State Communications*, from Pergamon is the better choice. Companion titles might include *Solid-State Ionics* from Elsevier, the *Journal of Solid-State Chemistry* from Academic, and *Acta Crystallographica B Structural Science* from Munksgaard—all titles discussed in the chapter on inorganic chemistry. If, on the other hand, there is a straightforward physics emphasis, or strong ties to the metals and materials community, the *Journal of Physics: Condensed Matter* from the Institute of Physics is a better choice. This title resulted from a merger of the formerly separate (British) Physical Society metal and materials journal with its solid-state journal.

At least two other families of solid-state journals, not depicted in these figures for reasons of space, are well worthy of consideration. They are the ironically titled *Philosophical Magazine* from Taylor and Francis, and *Physica Status Solidi* from the historically East German publisher Akademie Verlag. The first title, *Philosophical Magazine* has ancient origins as a journal first of general science and then of general physics, more recently specializing in solid-state physics without abandoning its rather surprising name. It has three editions: one for "Defects and Mechanical Properties" (Section A); one for "Structural, Electronic, Optical, and Magnetic Properties" (Section B); and one for rapid preliminary communications (Letters Section). *Philosophical Magazine* is to be preferred over *Physica Status Solidi*, a journal which has historically been surprisingly competitive despite its former East German sponsorship. *Physica Status Solidi* has a section A for applied research and a section B for basic work. Its historical lure was also of two parts. First, Soviet Bloc scholars who could not get either permission or acceptance into Western journals habitually sent their best papers to this largely English language and highly cosmopolitan journal rather than to their respective foreign-language national titles. Second, a fair number of Western European scholars came to prefer *Physica Status Solidi* as a good second-choice outlet for their manuscripts. The future of *Physica Status Solidi*, however, may diminish under German reunification. There is now virtually no restraint on Eastern Europeans whatsoever in submitting to historically West German journals or even to American titles. A core of good papers from formerly socialist Eastern countries may well desert *Physica Status Solidi*, and it is doubtful that featuring only second-choice papers from the West is enough to keep *Physica Status Solidi* a high priority choice in financially strapped libraries.

Those solid-state collections that are not already part of an engineering library will want to have at least some of the dozens of electronics titles in which solid-state physics figures prominently. Three of the more fundamental titles include the *IEEE Journal of*

*Solid-State Circuitry*, the *IEEE Transactions on Electron Devices*, and *Solid-State Electronics*, a Pergamon title.

Two of the more specialized topics rising to prominence within solid-state physics include *Superlattices and Microstructures*, an Academic Press title, and Elsevier's *Molecular Crystals and Liquid Crystals*. The former title has substantial interest for those analyzing the atom-by-atom networking of solids and how variations affect properties. The latter title concerns a category of materials that is often not quite solid, but not a freely flowing liquid either. Indeed, the liquid crystal light-emitting diode is what we see as the face of many electronic watches, signs, and some lap-top computer screens. Indeed, the modest progress within this liquid area is part of the reason (along with the merge-in of metals and materials papers) that a number of journals formerly entitled or subtitled "solid-state" have switched to "condensed matter," since liquids, like solids are condensed relative to gases and most plasmas.

Reviews in solid-state physics are somewhat more plentiful than reviews in applied physics overall. CRC Press publishes *Critical Reviews in Solid-State and Materials Sciences*, a softbound quarterly, while Academic publishes *Solid-State Physics*, a hardbound irregular. Both are recommended for virtually all solid-state collections.

## The Flush of New Journals in Superconductivity

At least four journals, too new to evaluate extensively, have arisen in superconductivity studies. They undoubtedly will be soon joined by others. A key issue is whether superconductivity scientists, whose field is so hot now and who have ready access not only to the best applied physics and solid-state journals but also to the pages of elite multiscience journals, will send their best papers to the new subspecialty titles. Nonetheless, any institution where superconductivity is being seriously pursued ought to have at least a few of these new titles, to hedge their bets.

Probably the best established title is *Physica C Superconductivity*. To avoid confusion, it's best to note that this journal is *not* related to *Physica Status Solidi A or B*. Rather, it is part of a family of Elsevier specialty physics journals that run from *Physica A* to

*Physica D*. Most of these *Physica* sections have a more modest standing relative to some of the other physics specialty journals discussed thus far in the book. Since a few hundred libraries within the U.S. had already taken advantage of Elsevier's reduced rate for taking all of these sections, section "C" for superconductivity began its career with a well-developed distribution system. Currently *Physica C* clearly publishes more superconductivity papers than any of its competitors. Unfortunately, this advantage is dimmed somewhat by its costing about four times as much when subscribed to singly. Nonetheless, it provides the best view thus far on Continental European work in the field. Most curiously, Elsevier has very recently launched an in-house competitor, *Superconductivity Theory and Applications*, based at SUNY Buffalo, presumably to attract U.S. papers. This seems to be a substitute for working harder to increase the appeal of *Physica C* to American contributors. It is not clear that this will help Elsevier, so much as two of Elsevier's competitors.

Those two competitors already have advantages in price and in other areas. Plenum, a publisher with a number of strong titles, has begun the *Journal of Superconductivity*. The British-based Institute of Physics is now publishing *Superconductor Science and Technology*. The latter title's distribution is being handled within the U.S. by the American Institute of Physics, with all the advantages of association and advertising that entails. Neither of these journals shows any signs of being below the standards of *Physica C* or, indeed, of their respective publisher's impressive inventory. The situation is very fluid and quite competitive. The Plenum title probably represents the best bet just now.

### Journals of Materials Science

While superconductivity is certainly the glamour field within solid-state physics, faster economic returns are much more likely in other specially tailored materials. The smart money poured into this seemingly more mundane effort is being partly absorbed by an increasing variety of materials science journals. Figures 5 and 6 are good indicators of the journal situation up to 1990 in the sense that

FIGURE 5

# Contributors and Relative Impact
# Leading Materials Science Journals

■ U.S.　▨ OTHER MAJOR POWERS　☐ IMPACT FACTOR

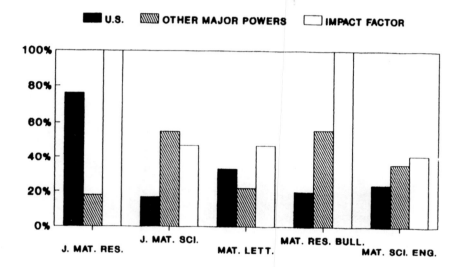

they stress the leaders in materials science from its roots in traditional engineering specialties.

Not surprisingly, the U.S.-based *Journal of Materials Research*, from the Materials Research Society, leads the pack. It is however, strongly pressed by an international entry from Pergamon, the *Materials Research Bulletin*. The surprise here is the relative affordability of the *MRB*.

The foreign press also features two competing families of journals. Both families are recommended in large collections, but some distinctions can be made that aid selection when there is less money.

For a number of years, Chapman and Hall, a small but reputable British house, has offered combined subscriptions to the *Journal of Materials Science* and the *Journal of Materials Science Letters*. Together these two journals provide the greatest number of papers,

FIGURE 6

## Comparison of Number of Papers, Costs, and U.S. Market Penetration

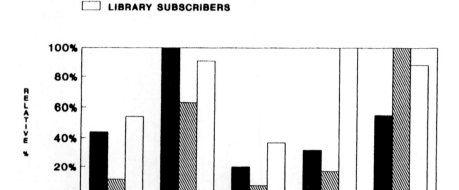

and they are a surprisingly less expensive purchase than their European competitors from Elsevier. The Chapman and Hall family is therefore recommended for the smallest collections.

However, Elsevier's family of materials titles has had advantages in depth of coverage that have made them quite competitive. Elsevier's family leaders are the combined *Materials Science and Engineering,* A and B sections. There are separate editions for "Structural Materials: Properties, Microstructure, and Processing" (A), and for "Solid-State Materials for Advanced Technology" (B). Elsevier's *Materials Letters* is approximately a match for the *Journal of Materials Science Letters*. But then Elsevier's advantage really begins to show with *Materials Chemistry and Physics* — not shown in the figures — with its strong ties to the chemical community, particularly to the solid-state inorganic chemists that are moving aggressively into materials research.

Two more, and very recent, entries are further evidence of the heightened interest in materials research on the part of the chemical community that Elsevier was first to detect. The American Chemical Society started the *Chemistry of Materials* in 1989, and during 1990 VCH, the distinguished German publisher will spin off into its own journal, *Advanced Materials*, formerly a section of its renowned *Angewandte Chemie - International Edition in English*. Within a year or two these new titles will have to be compared against the strong group already depicted in the figures.

Chapman and Hall, responding to the explosion of special interest materials titles, and seeking to keep its flagship *Journal of Materials Science* titles ahead in the race with Elsevier, has just initiated *Journal of Materials Science - Materials in Medicine* and *Journal of Materials Science - Materials in Electronics*. These new entries will also compete with a number of other better-established titles slanted towards these clients for materials science. The *Journal of Biomedical Materials Research* from Wiley and *Biomaterials* from Butterworth are two prominent examples in the first instance. The *Journal of Electronic Materials* from the Minerals, Metals, and Materials Society is an example in the second.

Other important specialized titles include the *Journal of Crystal Growth* from Elsevier, a title with a history of past emphasis on crystals grown from "wet" solutions but with a strong current emphasis on crystals grown from molten materials for the semiconductor industry. Likewise, Elsevier's *Journal of Magnetism and Magnetic Materials* is critical for many electrical and electronics collections.

*JOM* is a broader focus journal of the Minerals, Metals, and Materials Society with an origin as *JOM: Journal of Metals*. (It wisely chose to remain *JOM* rather than switch to *JO3M*, which would confuse it with the makers of Scotch tape, and it avoided *JOMMM*, which would have sounded like a mantra!) *JOM* is still the best source for metals industry developments. The *Journal of the American Ceramic Society* is the leading organ for that industry. While there are a great many polymer titles, fewer than expected emphasize interactions with the rest of the materials community. The *Journal of Composites Technology and Research* from the American Society for Testing and Materials, the *Journal of Composite*

*Materials* from Technomic, and *Polymer Composites* from the Society of Plastics Engineers are the best titles. *Synthetic Metals* is a special case in that some of its papers involve composites that are based on carbon, but the resulting materials are not generally polymers or "organic" in the commonly understood chemistry sense. The carbon materials most often involved are more like graphite or diamonds, which are among the less frequently discussed "inorganic" forms of carbon. The journal *Carbon* from Pergamon also serves this small but important community of carbon producers and users.

Review journals of materials science are led by the *Annual Review of Materials Science*, from Annual Reviews, Inc. CRC Press publishes *Critical Reviews in Solid-State and Materials Science*, while Elsevier issues *Material Science Reports*. A final title is *Progress in Materials Science*, from Pergamon. This author recommends selection in the order listed here.

### Journals of Surface Science

A key factor influencing choice of surface science titles is the degree to which the subscriber's agenda is tied to the preparation and analysis of electronic wafer materials, particularly as processed in vacuum conditions. The titles in Figures 7 and 8 are divided between the more general titles and those with a definite solid-state and vacuum emphasis.

Two general surface science titles should be in all collections. One is *Langmuir*, from the American Chemical Society. It has been previously discussed in the chapter on chemical physics, and is not repeated in the figures here. The other required title is *Surface Science* from Elsevier, depicted in the figures. Despite its price, it should be regarded as the godfather of all other surface science journals.

The remaining choices for collections heavily committed to electronic materials are also clear, and dominate these figures to the exclusion of the more chemically oriented titles. *Thin Solid Films* from Elsevier is absolutely key, if not inexpensive. Some savings will be realized by the relatively modest costs of the *Journal of Vacuum Science and Technology* sections *A* and *B*, (American Insti-

FIGURE 7

# Contributors and Relative Impact
# Leading Surface Science Journals

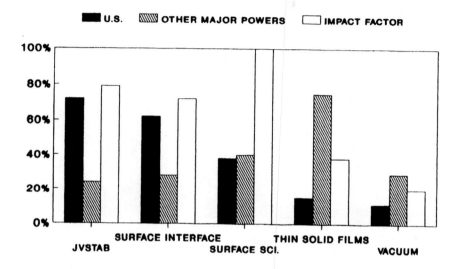

tute of Physics) and *Vacuum* (Pergamon). All of these journals feature studies on the treatment of surfaces with plasmas and ion beams. They are at the cutting edge of the basic science underlying today's chip technology. Sequential purchases in the order listed are recommended.

*Surface and Interface Science* from Wiley is one of two recommended final choices in surface science. The other is *Applied Surface Science*, another Elsevier title, not depicted in these figures, but similar in characteristics and strengths to its parent journal *Surface Science*. A choice may be made between them in that the Wiley title is less devoted to electronic materials, while offering some coverage on that subject.

A chemically oriented surface science collection would require some titles not depicted in the figures. Particularly recommended are the *Journal of the Electrochemical Society*, the *Journal of Elec-*

FIGURE 8

## Comparison of Number of Papers, Costs, and U.S. Market Penetration

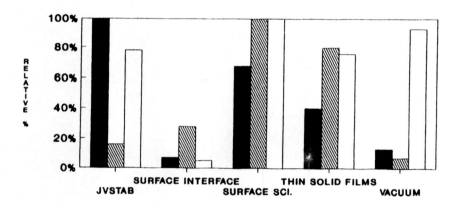

*troanalytical Chemistry and Interfacial Science* (Elsevier) and *Electrochimica Acta* (Pergamon).

Facilities with low energy accelerators and an interest in sputtering or other beam work would definitely need *Nuclear Instruments and Methods in Physics Research: Beam Interactions with Materials and Atoms* from Elsevier. Facilities which use X-ray probes of surfaces—a common technique—will probably require the *Journal of Electron Spectroscopy* from Elsevier. A new title, *X-Ray Science and Technology*, from Academic is also worthy of consideration.

Facilities that use electron microscopy to analyze surfaces should be careful of their journal choices. There is a strong tendency for given "ultrastructure" journals—as electron microscopy serials are frequently termed—to generally favor either physical science or life science topics. *Scanning*, a title from FACM, Inc., has the most appropriate emphasis for the surface science audience, followed by *Ultramicroscopy* from Elsevier.

There are few review journals devoted to surface science. Pergamon's *Progress in Surface Science* and Elsevier's *Surface Science Reports* are the best known examples.

### Journals of Optics

The journals discussed in this section are the most general titles in a field with many narrow interest technical titles. While virtually all optics collections will take each title mentioned in Figures 9 and 10, the selection of the special interest titles will be highly dependent on local activity.

The Optical Society of America, through its publishing link with the American Institute of Physics, absolutely dominates the broader interest optical journals. *Applied Optics*, the *Journal of the Optical*

FIGURE 9

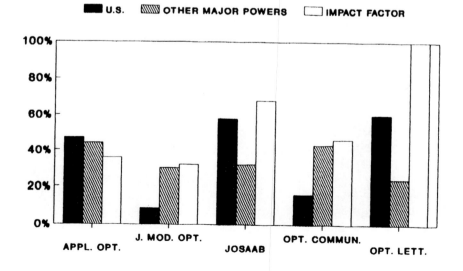

## Contributors and Relative Impact
## Leading Optics Journals

FIGURE 10

## Comparison of Number of Papers, Costs, and U.S. Market Penetration

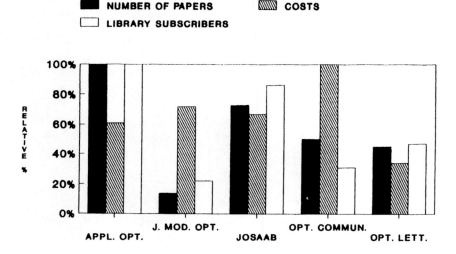

*Society of America*, sections *A* and *B*, and *Optics Letters* are all published by this group, and share many qualities and format features. There are some distinctions worth noting, however. Both traditional geometric optics and the latest ideas on image storage, enhancement, and retrieval are to be found in section A of *JOSA*, *Optics and Image Science*. *JOSA B: Optical Physics* is much more devoted to the results of spectroscopic investigations, and the fundamental characterization of the chemicals or surfaces probed is more important than the optical methods and tools used. On the other hand, methods and tools, and, in particular, remote sensing and scanning devices, are stressed in *Applied Optics*. With no special subject biases *Optics Letters* is truly general, but does specialize in brief, quickly published papers. A certain sorting is possible in light of these nuances. If your optics community is largely based on engineering, then *JOSA A* and *Applied Optics* are the best bets.

If there is a strong tie-in with atomic or molecular physics departments, then *JOSA B* is the key title. Everyone should take *Optics Letters*.

The two remaining titles represent some of the better foreign entries. Elsevier's *Optics Communications* is not, however, particularly devoted to fiber optics communications. It is a rapid communications journal in the style of *Optics Letters*. The *Journal of Modern Optics* is the latest reincarnation of a reputable Continental title, *Optica Acta*, that fell on hard times. This now highly functional Taylor and Francis journal gets the nod over a highly respected but rather traditional entry, *Optik* (Wissenschaftliche Verlag), which still publishes a few papers in German.

The leading review journal in the field is *Progress in Optics*, a hardbound series from Elsevier.

## Journals of Lasers and Fiber Optics

The two hottest areas of applied optics, lasers and fiber optics, have each generated new serials with some overlap, understandable given their interdependence.

Some evidence for the importance of lasers is that the leading trade magazine, *Laser Focus World* from Penn Well, has fifty-four thousand subscribers! It is certainly recommended for most engineering collections. (It has several competitors with similar levels of support.) The *Journal of Laser Applications*, from the Laser Institute of America is somewhat more scholarly than this semipopular genre, and is also recommended.

The *IEEE Journal of Quantum Electronics* is, however, the U.S. leader in serious laser research, while *Laser and Particle Beams* from Cambridge University Press represents some of the best U.K. and Commonwealth research. The *Soviet Journal of Quantum Electronics*, a translation by the American Institute of Physics of the leading Russian title, should be in all large collections, given the sustained interest the Soviets have shown in this field.

Work from Western Europe and Japan predominates in Springer's *Applied Physics B - Photophysics and Laser Chemistry*. In a similar vein, Harwood's *Laser Chemistry* deals not only with the composition of laser core materials, but with the especially sharp chemical kinetics and spectroscopy made possible by lasers. (The latter two titles are discussed in the section on chemical physics.) Both the higher intensity laser and the laser as a robotic scanner in industrial applications are featured in *Optics and Lasers in Engineering* from Elsevier, and in *Optics and Laser Technology* from Butterworth. Clinical journals involving lasers are discussed in the section on medical physics.

The *Journal of Lightwave Technology* from the IEEE is important to the straightforward laser community and absolutely key to its fiber optics branch. Likewise, a highly unusual journal — basically a somewhat loosely structured series of several dozen workshops and conference reports annually — entitled the *Proceedings of the SPIE* — is also quite important for both laser and fiber optics concerns. It comes from the Society of Photooptical Instrumentation Engineers, and is particularly recommended where high-tech instrumentation or electronic sensor and actuator work is a company interest.

There are a number of other titles in fiber optics, many are quasi-technical. Two are particularly noteworthy. One favors engineering overviews and includes fiber optics business surveys as well. The other title is more purely technical. The first is *Fiber and Integrated Optics* from Taylor and Francis. The second, more exclusively scientific title, is *Journal of Optical Communication* from Schiele and Schoen, a German firm. Neither of these sound titles, however, commands quite as many U.S. laboratory papers as the *IEEE Transactions on Communications* and its partner *IEEE Journal on Selected Areas of Communications*. Along with the aforementioned *Journal of Lightwave Technology* they represent the first round of choices in fiber optics literature. If a library can take only one of the non-IEEE journals, the Taylor and Francis title should get the nod, given its review journal function and despite its less technical nature.

Most engineering collections will want to take some of the corporate "house" journals featuring fiber optics progress in given companies. The best known example is the *AT&T Technical Journal*, although most of the larger U.S., European, and Japanese firms publish their own versions. The leading fiber optics trade journal is probably *Lightwave*, which is, not surprisingly, from Penn Well, the publishers of the leading laser magazine.

Those optical scientists edging closer to robotics and the perceptual psychology of vision would be well served by Pergamon's *Pattern Recognition* and Academic's *Computer Vision, Graphics, and Image Processing*.

## Journals of Acoustics

The first choice among acoustics journals is clear. (See Figures 11 and 12.) The *Journal of the Acoustical Society of America*, from the American Institute of Physics, is the world leader in most phases of the field, and is particularly strong in underwater sound. (Submarine detection is by far the most generously funded area of acoustics work, even in this peaceful era.) However, another title, Academic's *Journal of Sound and Vibration*, leads in certain broad, traditional, industrial aspects of acoustics, particularly vibration, metal fatigue detection, turbulence in fluid flow, and the damping

FIGURE 11

# Contributors and Relative Impact
# Leading Acoustics Journals

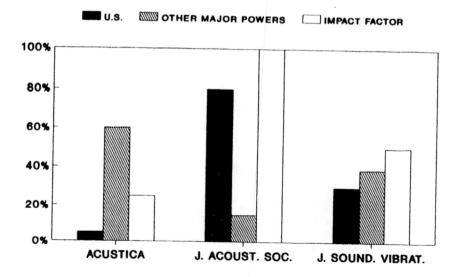

FIGURE 12

## Comparison of Number of Papers, Costs, and U.S. Market Penetration

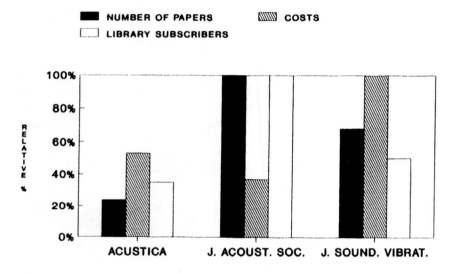

down of unwanted noise. This journal is mandatory in technical institutes with any interest in structural or mechanical engineering. Such collections will also want to have the even narrower, more machine-design-oriented *Journal of Vibration, Acoustics, Stress, and Reliability in Design* from the American Society of Mechanical Engineers. *Noise Control Engineering Journal* from Auburn University is yet another self-explanatory title of merit. *Acustica* from Hirzel, a small German publisher, has a wider range of acoustics interests than the Academic entry, but has not attracted many papers from American authors, despite its broader agenda. It remains a good third choice.

The special interests within acoustics have given rise to their own journals. The most notable examples have been in ultrasonics, which, as an offspring, is almost as large as its acoustical parent now. For physical sciences and engineering collections, the best ultrasound journals are a trio: *Ultrasonic Imaging* from Academic, *Ultrasonics* from Butterworth, and the *IEEE Transactions on Ultrasonics, Ferroelectrics, and Frequency Control*. The latter title is an excellent example of the integration of acoustics with solid-state physics and computing, as is its cousin, the *IEEE Transactions on Signal Processing*.

Other aspects of acoustics, including hearing research and audiology, might well find a place in a general acoustics collection. The most noted basic science journals in this regard are named, appropriately enough, *Hearing Research* from Elsevier and *Audiology* from Karger. The "broadcast and sound quality" constituency of acoustics is represented by the *Journal of the Audio Engineering Society*.

There are few reviews in acoustics outside occasional papers in the regular research journals.

## Journals of Medical Physics

Journals of medical physics differ in society sponsorship and in emphasis on items within the general topic of medical physics. Each society pays some attention to the subject matter favored by its competitors, but retains a characteristic preponderance. (See Figures 13 and 14.)

*Medical Physics*, from the American Association of Physicists in Medicine, focuses on the precise delivery of therapeutic radiation. Papers usually involve powerful beam instruments that require much computer-guidance in aiming. Curiously, calibration is done with mannequins called "phantoms." These phantoms, which are frequently pictured in articles and ads, contain internal detectors which sense the amount of radiations delivered at given instrument settings. *Medical Physics*, however, has less information on the management of radioactive leaks or environmental monitoring. That is the strength of *Health Physics*, a journal which Pergamon handles on behalf of the U.S.-based Health Physics Society. *Health Physics* has less emphasis on beam therapies. Both *Medical Physics* and *Health Physics* are largely American authored, and represent first choices for general collections in applied physics.

A British-based journal issued with substantial international soci-

FIGURE 13

## Contributors and Relative Impact
## Leading Medical Physics Journals

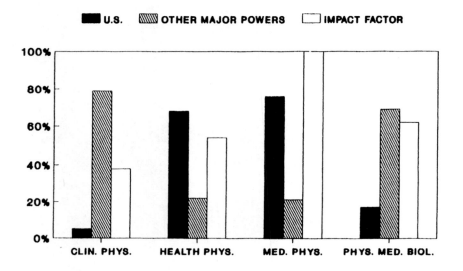

ety collaboration is *Physics in Medicine and Biology* from the Institute of Physics. It has a somewhat better subject balance than either of the two American entries, and includes a little more on medical imaging, the visual representation of internal body structures using less invasive probes. Nonetheless, it remains a better choice for audiences that favor *Medical Physics* over *Health Physics*.

The best coverage of a wide variety of procedures for visualizing a patient's internal structure and bodily function may be the relatively unsung *Clinical Physics and Physiological Measurement* from the British-based Institute of Physical Sciences in Medicine. This may be a better choice for collections that wish to be more well-rounded than even *Physics in Medicine and Biology. Clinical Physics* does have some fairly serious disadvantages in price-per-paper, but if this rather wide-ranging title attracted more subscriptions its cost might be more widely borne and thus stabilize.

FIGURE 14

## Comparison of Number of Papers, Costs, and U.S. Market Penetration

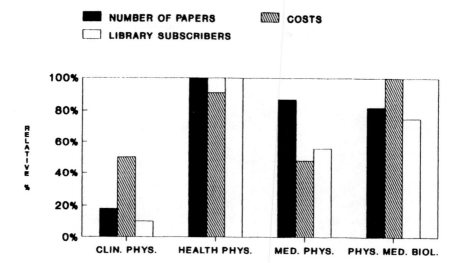

■ NUMBER OF PAPERS     ▨ COSTS
☐ LIBRARY SUBSCRIBERS

### Collateral Reading for the Medical vs. the Health Physics Constituencies

There is a plethora of physician-author radiology titles. A representative few should be in collections favoring devotees of *Medical Physics*. *Radiology*, from the Radiological Society of North America, and the *American Journal of Roentgenology*, a Williams and Wilkins entry named after the discover of X-Rays, are the best known journals reporting original research. A valuable hardbound continuing series with review articles and a theme format is *Radiologic Clinics of North America* from Saunders. It is highly recommended.

The *Journal of Computer Assisted Tomography* from Raven is indispensable for serious radiology collections, as is *Magnetic Resonance in Medicine* from Academic. Most large hospitals and university medical physics research centers now have both categories of advanced diagnostic machines and need both titles.

There is also a strong biomedical authorship among constituencies that favor *Health Physics*. *Radiation Research* from Academic is the leader, with the *International Journal of Radiation Biology* from Taylor and Francis following closely. A relatively new title, *Radiation and Environmental Physics* from Springer, is promising for larger collections, although, like its competitors, it is quite biological in tone. Pragmatic and closer to the physics training of this audience are *Radioactive Waste Management* from the U.S. Department of Energy and *Radioactive Waste Management and the Nuclear Fuel Cycle* from Harwood. Both of these titles are recommended for any atomic energy or nuclear weapons facility.

## Journals of Acoustics and Optics in Medicine

Given the importance of ultrasound, all three of the following titles should be in most medical physics collections. These are the *Journal of Clinical Ultrasound* from Wiley, the *Journal of Ultrasound in Medicine* from Elsevier, and *Ultrasound in Medicine and Biology* from Pergamon. Most of the authors will be physicians.

Optics in medicine also is covered largely by clinicians in *Lasers in Surgery and Medicine* from Wiley-Liss. Fiber optics is featured in *Endoscopy*, a Thieme title. Nonetheless, both journals deal with the interests of an important class of applied physics customers and are worth including in a large medical physics collection.

The reader may ask, are there any other titles of importance to medical physicists in which doctors do *not* predominate? The answer is yes, and hearkens to the Hounsfield tradition of electrical engineers making sophisticated instruments. Strongly recommended are the *IEEE Transactions on Medical Imaging* and the *IEEE Transactions on Biomedical Engineering*.

# Subject Index

Academic Press
  acoustics journal, 219-220
  analytical chemistry journals,
      23-24,30-31
  applied physics journals, 205,214
  inorganic chemistry journals, 48,
      52,53,205
  medical physics journals, 161,
      167,223,224
  organic chemistry journals, 78-79
  physical chemistry journals, 121,
      124-125
  solid-state physics journals, 207
  ultrasound journal, 221
Accelerators, nuclear, 151-154,156
  journals of, 164
Acid-base compounds, 38-40
Acoustics, 186-187,198-199
  journals of, 219-221,224
Adams, Roger, 67-68
Aerodynamics, 182,183
Aerosols, 115-116
Agricultural chemistry, journals of,
      29
Akademie Verlag, 206
Akademische Verlagsgesellschaft,
      118-119
Alchemy, 34
Alkaloids, 61,67
Allinger, Norman, 68-69
American Academy of Forensic
      Sciences, 31
American Association of Cereal
      Chemists, 29
American Association for Clinical
      Chemistry, 30

American Association of Physicists
      in Medicine, 221
American Astrophysical Union, 173
American Chemical Society, 19,29,
      44-45,47,74,79,85,116,212
  analytical chemistry journals, 19,
      29
  inorganic chemistry journals,
      44-45,47
  organic chemistry journals, 74,79
  physical chemistry journals, 117,
      123,128,131
  polymer chemistry journals, 85,86
American Institute of Physics
  acoustics journals, 219
  applied physics journals, 200
  chemical physics journals, 120,
      121-122
  nuclear physics journals, 163,164,
      165-166,167
  optics journals, 215-216,217
  plasma physics journals, 171
  superconductivity journals, 208
  surface science journals, 212-213
American Society of Mechanical
      Engineers, 220
American Pharmaceutical
      Association, 79
American Physical Society, 160,200
American Society
      for Pharmacognosy, 82
American Society for Photobiology,
      128
American Society for Testing
      and Materials, 29,211-212
Amino acids, 67

226    *MAKING SENSE OF JOURNALS IN THE PHYSICAL SCIENCES*

Analytical chemistry. *See*
    Chemistry, analytical
Analytical instrument
    manufacturing, journals of,
    29-30
Aniline dyes, 58
Annual Reviews, Inc., 120,164-165,
    212. *See also* Review
    journals
Antimatter, 143-144,155
Antiproton, 155
Antisemitism, 100-101,106
Arrhenius, Svante, 98-99,100
Association of Official Agricultural
    Chemists, 29
Aster Press, 27
Aston, Francis William, 28
Astrophysics, 158-159
    journals of, 173-175
Atomic bomb, 149-150,157
Atomic models, 102-104
    Bohr's, 102,140-141,143-144
    Einstein's, 143
    Rutherford's, 137-139
Atomic physics. *See* Physics, atomic
    particle
Atomic theory, 36,136-140
Atomizer, 116
Austria, 12

Baeyer, Adolf von, 58-59,65
Bakelite, 71
Bancroft, Wilder, 101
Bard, Allen J., 18
Bardeen, John, 183-184, 192
Barth, 44
Barton, Derek, 69,74
Basov, Nicolas, 185-186
Batavia, Illinois, nuclear accelerator,
    152-153
Becker, Herbert, 145
Becquerel, Henri, 136
Bednorz, J. Georg, 192,203
Belgium, organic chemistry in, 59,
    64

Bell Labs, 183-184,186
Bentley's compound, 67
Benzene, 63
Berzelius, Jakob, 59-60,61,93,97
Biochemistry, 65-66. *See also*
    Chemistry, organic
    pharmaceutical, 67
Biography, scientific, 1-3
Bioorganic chemistry, 42,65-66
    journals of, 55,78-79
Black, Joseph, 35
Blackwell, 29,173
Bloch, Felix, 28,194
Blodgett, Katherine 116
Bloembergen, Nicolaas, 111,186
Bohr, Aage, 141
Bohr, Niels, 102,140-141,142,
    143-144,146,148,149
Boltzmann, Ludwig, 95-96,100
Bose, S. N., 155
Boson, 155
Bothe, Walter, 145
Boyle, Robert, 36
Bragg, W. H., 49
Bragg, W. L., 49
Brattain, Walter, 183-184
British Institute of Physics, 170
British Royal Society of Chemistry,
    44,45
Brown, Herbert, 41
Buchner, Eduard, 65
Bunsen, Robert, 15,40,105,191
Bunsen burner, 105
Bunsen Society, 118
Butenandt, Adolf, 67
Butterworth
    chemical catalysis journals, 52
    polymer chemistry journals, 85-86
    ultrasound journals, 221

Cacodyl, 41,105
California, silicon chip industry, 188
California Institute of Technology,
    39
Cambridge University, 137

Microscale, 15
Microwaves, 185-186
Millikan, Robert, 39,40,137
Mine explosions, 115-116
Minerals, as inorganic substance
　source, 35-36
Miniaturization, electronic, 183-184,
　188,199
Moelwyn-Hughes, Emyr Alun, 113
Molecular orbital studies
　computer graphics in, 106-107
　historical background, 102-107
　journals of, 125-128
Morton-Thiokol, 72
MRI (magnetic resonance imaging),
　200
Muller, K. Alex, 192,203
Mulliken, Robert, 39-40,106,107
Munksgaard, 50,205
Mussolini, Benito, 146,148

Naphthalenes, 61,63
Napoleon Bonaparte, 90
National Institute of Standards
　and Technology, 186
National Research Council
　(Canada), 161
Natural products chemistry, 66-67,
　70-71
　journals of, 82-84
Nazism, 100-101,106,187
Nernst, Walther, 96,100,101,112,
　116
Netherlands, chemical
　instrumentation
　manufacturers, 12
Neutrino, 144,145-146,155
Neutron, 145-147
Newton, Isaac, 104,197
Nickon, Alex, 68
Nicotine, 61
Nieuwland, Fr., 73
Nitrogen, extraction from
　atmosphere, 96

Nobel Prize winners
　Arrhenius, Svante (1903), 100
　Aston, Francis W. (1922), 28
　Baeyer, Adolf von (1905), 58-59
　Bardeen, John (1956 and 1972),
　　183,192
　Barton, Derek (1969), 69
　Becquerel, Henri (1903), 136
　Bednorz, J. Georg (1987), 192
　Bloch, Felix (1952), 28
　Bloembergen, Nicolaas (1981),
　　111
　Bohr, Aage (1975), 141
　Bohr, Niels (1922), 141
　Bragg, W. H. (1915), 49
　Bragg, W. L. (1915), 49
　Brattain, Walter (1956), 183
　Brown, Herbert (1979), 41
　Buchner, Eduard (1907), 65
　Butenandt, Adolf (1939), 67
　Chamberlain, Owen (1959), 155
　Cherenkov, Pavel (1958), 158
　Cockcroft, John (1951), 151
　Cooper, Leon (1972), 192
　Cormack, Allan (1979), 190
　Curie, Marie and Pierre (1903),
　　136
　de Broglie, Louis (1929), 142
　Debye, Peter (1936), 100
　Eigen, Manfred (1967), 111
　Einstein, Albert (1921), 143
　Fermi, Enrico (1938), 148
　Feynmann, Richard (1965), 155
　Fischer, Emil (1902), 65
　Flory, Paul (1974), 74
　Fukui, Kenichi (1981), 104
　Grignard, Victor (1912), 41,62
　Haber, Fritz (1918), 100
　Hahn, Otto (1944), 43
　Hassel, Odd (1969), 69
　Hauptman, Herbert (1985), 43,49
　Haworth, Walter (1937), 67
　Heisenberg, Werner (1932), 143
　Herzberg, Gerhard (1971), 106

[Indexes compiled by Kathleen J. Patterson]

**RETURN TO** ➡ PHYSICS LIBRARY
351 LeConte Hall          642-3122

| LOAN PERIOD 1 1-MONTH | 2 | 3 |
|---|---|---|
| 4 | 5 | 6 |

ALL BOOKS MAY BE RECALLED AFTER 7 DAYS
Overdue books are subject to replacement bills

## DUE AS STAMPED BELOW

|  |  |  |
|---|---|---|
|  |  |  |
|  |  |  |
|  |  |  |
|  |  |  |
|  |  |  |
|  |  |  |
|  |  |  |
|  |  |  |
|  |  |  |
|  |  |  |
|  |  |  |
|  |  |  |
|  |  |  |

FORM NO. DD 25

UNIVERSITY OF CALIFORNIA, BERKELEY
BERKELEY, CA 94720

Ⓟs